First International Conference on Energy, Power, Petroleum and Petrochemical Engineering (E3PE 2017)

MATEC Web of Conferences Volume 171 (2018)

Beirut, Lebanon
26 – 28 April 2017

Editor:

Nazih Moubayed

ISBN: 978-1-5108-6389-7

Printed from e-media with permission by:

Curran Associates, Inc.
57 Morehouse Lane
Red Hook, NY 12571

Some format issues inherent in the e-media version may also appear in this print version.

Printed by Curran Associates, Inc. (2018)

For additional information, please contact EDP Sciences – Web of Conferences at the address below.

EDP Sciences – Web of Conferences
17, Avenue du Hoggar
Parc d'Activité de Courtabœuf
BP 112
F-91944 Les Ulis Cedex A
France

Phone: +33 (0) 1 69 18 75 75
Fax: +33 (0) 1 69 28 84 91

contact-edps@webofconferences.org

Additional copies of this publication are available from:

Curran Associates, Inc.
57 Morehouse Lane
Red Hook, NY 12571 USA
Phone: 845-758-0400
Fax: 845-758-2633
Email: curran@proceedings.com
Web: www.proceedings.com

TABLE OF CONTENTS

Preface

On behalf of The First International Conference on Energy, Power, Petroleum and Petrochemical Engineering (E3PE2017), the program committees of the conference, Faculty of Engineering – Lebanese University, Beirut, Lebanon, have the pleasure to welcome the attendees to Faculty of Engineering – Lebanese University, Beirut, Lebanon to participate in The First International Conference on Energy, Power, Petroleum and Petrochemical Engineering (E3PE2017) during the period of April 26-28, 2017. In this edition, E3PE2017 is providing a venue for high quality papers in a wide range of research areas in the energy, Power, Petroleum and Petrochemical Engineering fields. The conference has recorded the receipt of around 52 papers, out of which 18 were accepted forpublication, resulting an acceptance rate of 35%. The accepted papers were authored by researchers from more than 10 countries. Each paper was evaluated by a minimum of two reviewers. Additional reviews were also conducted, if necessary. Authors received reviewers' comments to be considered in the final manuscript of their papers. We're distinctly pleased to have distinguished invited speakers. They will offer their lecturers over keynote presentation sessions. The parallel sessions cover the following topics: Renewable Energy, Petroleum and Petrochemical Engineering, Energy Management Systems Chemical Industry Technologies and Products, Enhanced Oil Recovery, Ecology and Environmental Technologie and many more related Petroleum and Petrochemical Engineering topics.

General Chair
Prof. Nazih Moubayed, Lebanese University, Lebanon

E3PE2017 Program Committees

Honorary Chair
Rafic Younes, Lebanese University, Lebanon

General Chair
Nazih Moubayed, Lebanese University, Lebanon

Deputy Chair and Head of Organizing Committee
Abed Ellatif Samhat, Lebanese University, Lebanon
Jean Chamoun, Lebanese University, Lebanon

Publication Chair
Yoshiro Imai, Kagawa University, Japan

Organizing Committee
Lisa Diab, Lebanese University, Lebanon
Joyce Kheir, Lebanese University, Lebanon
Sorina Mortada, Lebanese University, Lebanon
Rayane Slim, Lebanese University, Lebanon
Pierre Rahmeh, Lebanese University, Lebanon
Khalil Khoury, Lebanese University, Lebanon
Carole Aghnatios, Lebanese University, Lebanon
Wael Hamd, Lebanese University, Lebanon
Hamza Magnier, Lebanese University, Lebanon

International Program Committee
Ali Abbas, The University of Sydney, Australia
Ali Assi, IEEE, Lebanon
Catalin Popescu, Petroleum-Gas University of PLOIEST, Romania
Francesco Aymerich, University of Cagliari, Italy
Fawzi Banat, The Petroleum Institute, UAE
Hadi Kanaan, IEEE, Lebanon
Henri El Zakhem, Balamand University, Lebanon
Hiba Al-Sheikh, IEEE, Lebanon
Imad Hajj Shehade, Technology and development director IRI, Lebanon
Juan Carlos Santamarina, KAUST University, Saudi Arabia
Monzer Mostapha, IEEE, Lebanon
Mourad Henchiri, University of Nizwa, Oman
Nabil Karami, IEEE, Lebanon
Nayef Mohamed Ghasem, United Arab Emirates University, UAE
Pierre Elkhoury, Director of the Lebanese center for energy conservation, Lebanon
Redhouane Henda, Laurentian University, Canada
Tilda Akiki, IEEE, Lebanon
Waail Mahmod Lafta, Griffith University, Australia
Walid Kamali, IEEE, Lebanon

Statement of Peer review

In submitting conference proceedings to *Web of Conferences*, the editors of the proceedings certify to the Publisher that

1. They adhere to its **Policy on Publishing Integrity** in order to safeguard good scientific practice in publishing.
2. All articles have been subjected to peer review administered by the proceedings editors.
3. Reviews have been conducted by expert referees, who have been requested to provide unbiased and constructive comments aimed, whenever possible, at improving the work.
4. Proceedings editors have taken all reasonable steps to ensure the quality of the materials they publish and their decision to accept or reject a paper for publication has been based only on the merits of the work and the relevance to the journal.

Title, date and place of the conference

The First International Conference on Energy, Power, Petroleum and Petrochemical Engineering (E3PE 2017)
April 26-28, 2017 Beirut, Lebanon

Proceedings editor(s):

Prof. Nazih Moubayed, Lebanese University, Lebanon

Date and editor's signature

18 May, 2018

MATEC Web of Conferences **171**, 01001 (2018)
E3PE 2017
https://doi.org/10.1051/matecconf/201817101001

Stand Alone PV House Design for Maximum Power Output in Sydney Australia Metropolitan Area

M. Nassereddine, J. Rizk, A. Hellany, M. Nagrial
School of Computing, Engineering & Mathematics
Western Sydney University
Penrith, Australia
mnassereddine@ieee.org

A. Elrafhi, Z. Obeid, K. Hajar
School of Engineering
University of Technology and applied Science
Tripoli, Lebanon
Ahmad.rafhi@ulf.edu.lb

***Abstract*— Nowadays, a strong link exists between human comfort and electrical power. Abundant doings are not possible without the support of the electrical energy. Driven by the human obligation to save the environment, the installation of PV solar system increased dramatically in the past numerous years. The installation can be grid connected or standalone system. These micro-grid situations are installed world-wide and currently are contributing to the local electrical network. To maximize the use of the generated solar system, numerous system installed energy storage in the form of batteries. These energy storages harvest the sun power during daylight to support the household electrical demand after sunset. This paper linked the house roof design to the PV solar system for maximum generation condition. Furthermore the paper contains important information on the optimum location for the solar panel to ensure continuous and reliable system is achieved. An actual real case study is presented.**

***Index Terms*—Electrical Power, Energy Storage, House Consumption, PV System.**

I. INTRODUCTION

The installation of solar PV system is accelerating worldwide. Numerous countries support financially the installations of small PV system for house or industrial use. For example, in Australia, the government provides rebates for installing PV solar system, grid connected or stands alone, on the roof of residential or industrial properties.

The micro-grid installation at the residential properties supports the local network by injecting the surplus power into the utility network. Taking into consideration the fact that PV micro-grid produces maximum power during the midday period on a sunny day. The house peak demands hours are located in the morning and late in the afternoon/evening [1, 2]. Based on the mentioned information, the micro-grid generated power doesn't support the house critical electrical requirements.

Electrical power storage is required to harvest the sun power during the day to support the electrical apparatus during the night time. Due to the cost and life time of available energy storage, numerous researchers discussed numerous approaches to enhance the system efficiency and extend the light of the proposed storage system.

The paper provides information regarding the stand alone house requirements. The assessment includes numerous approaches on how to design an advanced architect of an efficient PV system with storage facility. Furthermore, the paper shows how the house building design can play a vital role to ensure maximum power is harvested from the sun. The presented case study is based on the constructed and commissioned standalone PV house in Sydney Australia metropolitan area.

II. THEORETICAL STUDY

Driven by the greenhouse effect and United Nation Goal 7 [3], the installation of the PV solar system for residential properties increased in the past decades. PV panels generated power changes throughout the year. Figures 1 and 2 show the average hourly solar distribution, and the electrical loading for residential customers [1, 2]. It is clearly shown that the generated power requires energy storage to support the house daily (24 hours period) electrical consumptions.

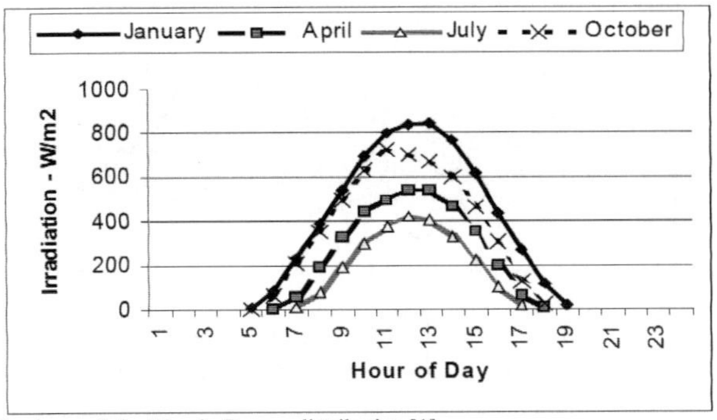

Fig. 1. Solar Hourly Energy distribution [1]

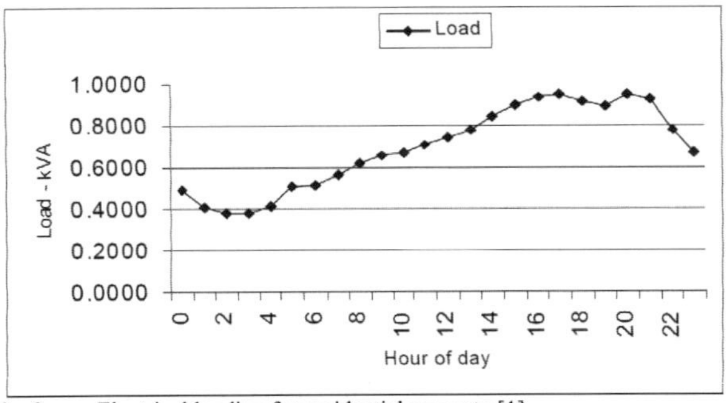

Fig. 2. Electrical loading for residential property [1]

The energy generated by the solar panel during the sun-light period supports the running electrical appliances and charge the batteries to support the electrical demand during the sun-set period. Based on this information, the size of the PV solar system and battery storage is critical to ensure system reliability and availability.

The standalone design in this paper is divided into the following sections:

- House orientation and roof layout
- Solar panels system sizing and layout
- Energy storage sizing
- House electrical distribution circuits

A. House Orientation and Roof Layout

The house orientation, roof space and tilt play an important role for the PV solar system. The PV solar panel maximum output power is obtained if the panel is installed perpendicular to the sun light. This can only be achieved if the roof supports such installation. For example, in Sydney, the sun path is shown in fig. 3. The figure also highlights the inadequate sun light exposure for PV panels facing the south. Furthermore, the figure shows that the sun angle to the panel changes, the angle depends on the seasons. For example, in spring, the optimum tilt angle is 34 degree while in summer is 16 degrees.

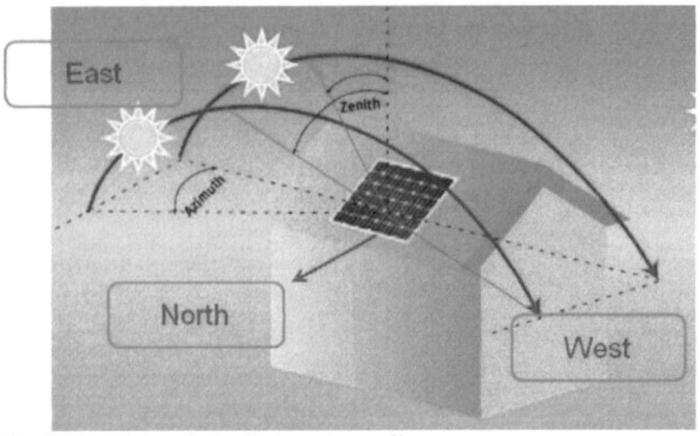

Fig. 3. Sun path for Sydney Australia

Based on the above information, the roof design (tilt and space) plays an important role for the house maximum possible power generation. The house electrical appliances and its operating sequence form one of the main inputs for the roof design. Take Sydney house as an example, if the critical house electrical demands is for the afternoon during the winter, the roof design should allow for sufficient space that is facing North-West with 58 degrees tilt angle. The proposed equation 1 in this paper can be used to assist with the roof space design and tilt.

$$\begin{cases} R_S = \left(\dfrac{H_{C-P}}{PV_P \times H_E} \right) \times A_P \\ T_A = \begin{cases} 16^0 - Summer \\ 34^0 - spring / Autumn \\ 58^0 - W\,\text{int}\,er \end{cases} \end{cases} \quad (1)$$

Where

R_S is the roof area requirements to support the required critical power during a defined day period

H_{C-P} is the house required critical power during a defined day period

H_E average daily sun exposed hours

PV_P is the single PV panel rated output power (including the efficiency coefficient)

A_P is the single PV area.

T_A is the required tilt angle for the roof and is based on the season that reflect the critical power requirements

B. Solar panels Sizing and Layout

The sizing of the PV panel system depends on the house electrical appliances and its operating pattern. The design should take into consideration the instantaneous power P_i and consumed power P_C. The paper used equation 2 [4] to compute the electrical power disbursed by the house electrical appliances:

$$P_C = K + O + L + G + M + H \quad (2)$$

Where

P_C is the total daily power consumed by the house (Wh)
K is the air-condition power (Wh)
H is the hot water power (Wh)
O is the cooking power (Wh)
L is the lighting power (Wh)
M is the washing and cleaning machines power (Wh)
G is the general power (Wh) (Fridge, hair dryer, mixer….)

The instantaneous power is based on the proposed operation sequence of the electrical appliances. The paper proposed the use of equation 3

$$\begin{cases} E_L = Equipment - List - Operating - same - time \\ P_i = \sum E_L \end{cases} \quad (3)$$

Where

E_L is each apparatus rated power (kW)

For example, if M, K and L are running at the same time, $P_i = M_i + K_i + L_i$

Where

$M_i, K_i \,\&\, L_i$ are the instantaneous power of each apparatus in kW

It is worth mentioning, the instantaneous power set the requirements for the charger inverter, PV panels and battery storage. While the consumed power set the requirements for the energy storage. The installed system should be able to support the instantaneous power which means the inverter instantaneous power should match/exceed the house requirements. The inverter instantaneous power is driven by the PV panels and battery instantaneous power capability during the day and the battery instantaneous power capability during the sun-set and heavily cloudy days.

The number of panels is determined based on the consumed power and the number of sun hours. Equation 4 can be used to compute the number of PV panels:

$$N_P = \frac{P_C}{PV_P \times N_H} \tag{4}$$

Where

N_P is the number of minimum panels to support the required electrical load

N_H is the number of sun hours that the panels will be exposed per day

It is worth mentioning, the determination of the electrical power usage per apparatus might require the manufacturer assistance. For example, if a 2.2kW dishwasher running for 2 hours, this doesn't mean it consumes 4.4kWh (more evidences are provided within the case study). Based on this information and to avoid over design (which cost money and space), it is worth requesting the consumption from the manufacturer for a new house or run a test on the existing house appliances.

The location and tilts of the panels determine the level of outputs for a given time of the day and year. The hourly loading consumption chart determines which part of the day the sun energy is critical to ensure continuous and reliable power is available. In addition, this process can also be used to extend the battery storage life cycle. Additional information is provided in the next section.

C. Energy Storage

Based on figures 1 & 2, maximum power demands in the afternoon are not supported by the PV solar generated power. To ensure continuous and reliable of the power supply is provided, energy storage is required as an essential element for the stand alone house. The concept is to store the excessive energy generated during the daylight to support the electrical load after sun-set. Based on this information, the optimum goal is to ensure the battery storage is fully charged by the sun-set or close to it. This provides the house with maximum power storage for the use overnights and the early mornings.

To ensure batteries are fully charged in the evening, the PV panels set up should provide sufficient power in the afternoon to support the house electrical consumptions and ensure the batteries maintain its fully charged power. In addition, fig. 2 shows that the maximum house electrical usage is located in the evening and early hours of the night. Therefore, PV panels facing toward the west support the afternoon power requirements better than the one facing east or north.

Sizing and life of the batteries depends on the following factors:

1. Number of back up days (during these days the battery should support the house electrical demands)
2. Electrical usage after sun-set.

1) Number of back up days

Equation 5 is used in this paper to compute the required battery size to ensure continuous electrical power during the backup days.

$$B_S = \frac{N_D \times P_C}{\zeta_B} \tag{5}$$

Where

B_S is the required battery storage (kWh)

N_D is the required backup days

ζ_B is the battery level of discharge (must be less than 1)

2) Electrical usage after sun-set

Batteries life is measured with cycle, each time the battery is fully discharged, and one life cycle is lost. The house consumptions after sunset (or during heavy cloudy day) utilizes the battery stored energy and start the life cycle clock. Reducing the energy consumption after sunset extends the battery life. This can be achieved by using advance energy management system. More information can be found in the author published paper [4].

D. House Distribution circuit

The house distribution circuit plays an important role to improve system efficiency for standalone houses. Dual electrical bus, direct Current (DC) and Alternating Current (AC) offers higher efficiency system (eliminate the energy requirement to perform the inversion when use DC appliances rated at the system voltage). Numerous household items, such as lighting, can operate on DC power. Fig. 4 represents the paper proposed standalone house distribution circuit.

The advantages of the DC bus can be presented in the following items:

- Removal of the inverter which reduces the losses and the budget

 https://doi.org/10.1051/matecconf/201817101001

- Choose appliances with operating voltage similar to the battery storage to eliminate the usage of DC/DC converter
- Design the PV array voltage to comply with the acceptable operating limits of the DC apparatus. This reduces the stress on the charger controller and extends its life.
- During sunny clear days, it is possible to run the DC appliances direct from the PV panels
- The charger controllers are cheaper than inverters

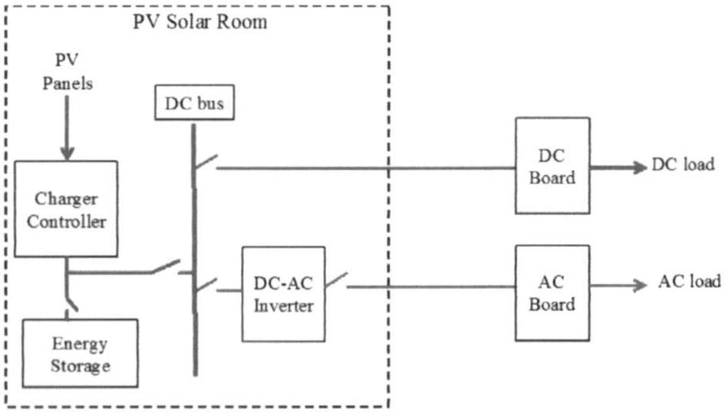

Fig. 4. Dual bus distribution circuit for the standalone house

III. Theoretical Study Discussion

The theoretical section of this paper highlights the important of the house roof architect design to maximum the sunlight exposure and to suit the timeline critical power demand. This approach supports the following discussion:

- Currently the battery storage cost is high in comparison to the PV panels price
- The battery storage operating life is short in comparison to PV panels.
- The installation of additional PV solar panels allow for:
  - Faster battery charging
  - Support the house hold during cloudy day without the aid of the battery
  - Reduce the battery size for backup days especially in Sydney Australia (It is rare to have a total dark cloudy days in Sydney Australia, Always will be exposure to sunlight (high or low density) during the day

The following example supports the above discussion: if a 10kW solar panels are installed to support the household with 20kWh battery storage, during a cloudy day, if the 10kW panels are operating at 20% for two hours during the day (low sunlight density), the PV panel produces 4kWh during the cloudy day which represents 20% of the battery storage capacity. If only 5 kW solar panels is installed, the output power during the same day is 2kWh which represents only 10% of the battery storage

In addition to the above, the theoretical study introduces equation 1 which assists with the roof design to maximize the power output. Also the analysis captures the orientation of the solar PV panels to support the critical power demands and maximize the use of the battery storage.

IV. Case Study

It is worth mentioning that the case study is based on the design and constructed house in Western Sydney region. The following form the input of the case study:

- The roof foot print of the proposed new house is 13 by 10 meters
- The land orientation is shown in fig. 5
- House electrical appliances:
  - 76 Down Light (DC LED Light)
  - Television (LED)
  - Refrigerator (230V 420kWh/year)
  - Freezer (230V, 360kWh/year)
  - Ducted Air-condition (single-phase 230V, 28A rated current)
  - DC Split Air-condition (48V, 1.2kw)
  - Electric Oven (2.2kW)
  - Electric Cooktop (2kW)
  - Hot water-Heat pump (700W)
  - Washing Machine (450W)
  - Microwave Oven (2.2kW)
  - Kettle (2kW)
  - Hoover (2.2kW)
  - Dishwasher (2.2kW)
- Battery level of discharge is 70%
- Summer average sun hour exposure is 6
- Winter average sun hour exposure is 4
- 250W PV panels to be used
- 1.95m^2 the area for single panel

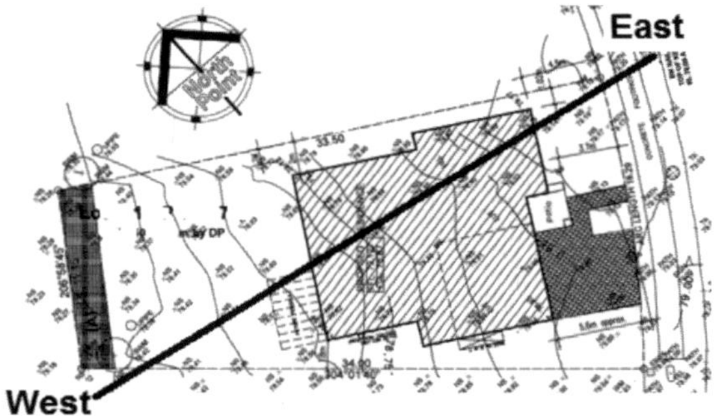

Fig. 5. House orientation in respect to the North

The roof design was an essential decision to ensure the required PV panels are installed to support the house electrical load. The assessment shows that in the maximum daily power consumption is located in the late afternoon period during the summer season. Based on this assessment, the house should have sufficient area to install panels that faces North West. The published works in [4] are used to compute the hourly electrical consumption for the house. It is worth mentioning the following two items as the experimental results supporting the theoretical statement:

- The dishwasher usually runs for 2.2 hours, therefore, based on the 2.2kW rated power, the consumption is computed to be 4.84kWh (the computation is completed by multiplying the rated power by the number of hours)
- The Oven usually runs for 0.7 hour, the computed consumption power is 1.54kWh

In Sydney Australia, there is no adequate sun exposure for the roof that is tilted and facing south. Therefore, the roof tilt should avoid the south. Fig. 6 shows the proposed roof design along with the tilt of the PV panels.

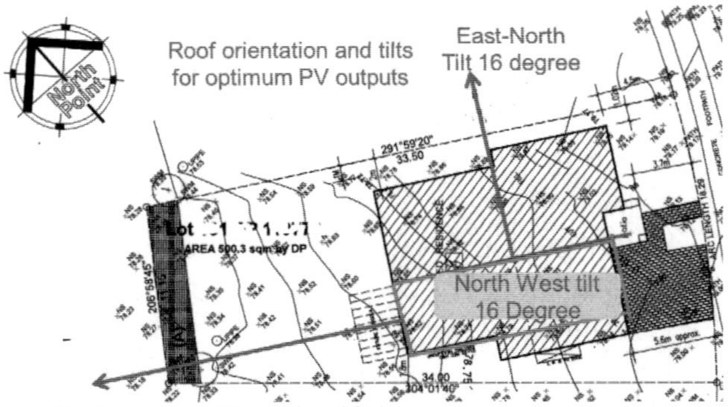

Fig. 6. Proposed roof design

Table 1 summarizes the daily electrical maximum loading with and without the Air-condition. It is worth mentioning that the hot water operates only 4 hours per day.

TABLE I. TABLE1: COMPUTED DAILY ELECTRICAL CONSUMPTION

Details	Morning (kWh) (06:00-10:00)	Midday (kWh) (10:00-14:00)	Afternoon (kWh) (14:00-18:00)	Evening and early night (kWh) (18:00-22:00)	Overnight (kWh) (22:00-06:00
Electrical consumption with Ducted Air-condition	3.2	16.1 (60% is after 12)	10.6	3.8	2.8
Electrical Consumption without Ducted Air-condition	3.2	2.9	2.1	2.6	2.8

Based on the table 1 data, the maximum daily power consumption with the ducted air-condition is 36.5kWh and the maximum daily power consumption without the ducted air-condition is 13.6kWh. The ducted air-condition will only operate during hot days, which mean the sun is shinny and the PV solar panels generating electrical power. The house PV system is designed based on the followings:

- The PV solar system produces sufficient instantaneous power during sunlight to support the house electrical demands with the presence of the ducted air-condition
- The battery storage is sized to provide two days backup support without the ducted air-condition

Equation one shows an area of 47.45m2 is required to install the required 250W panels. This equate to 28 panels.

$$\begin{cases} R_S = \left(\dfrac{36.5}{0.25\times 6}\right)\times 1.95 = 47.45m^2 \\ T_A = \{16^0 - Summer \end{cases}$$

Table 1 data shows that large consumption is located between 10:00 and 14:00. To ensure the PV output is staggered throughout the day, the system is divided into two sections:

- One facing North East
- One facing North West

It should be noted that the roof design has an area of 90m^2 available for the panel installation. Taking into consideration the theoretical discussion, the decision was made to install an additional 12 panels which make it's a total of 40 panels that equate to 10kW system.

Equation 5 is used to size the required battery storage:

$$B_S = \frac{2\times 13.6}{0.7} = 48.85Wh$$

The design used 2V 800AH battery cell to create the 48V 38.4kWh battery bank. The system was installed at the designated property. An energy monitoring system is installed to record the system consumption. Figures 7 and 8 show the house daily maximum instantaneous power as well as the energy consumption. During the day, the system was closely monitored where the followings are observed:

- During the sunny clear day, the output of the PV panel supports the house electrical demand without the use of the energy stored in the battery
- In the evening, the system facing North-West produces around 70% more power to the one facing north east which ensures the house electrical demands is met with minimum battery supports.
- The recorded actual energy consumption for the dishwasher is 45% of the value (4.84kWh) stated during earlier assessment. The monitoring system shows that the dishwasher has numerous cycles where the power increased to 2.2kW for a limited time before it dropped to 0.2kW.
- Similar observation is witnessed for the electric oven

Fig. 9 shows the measured daily instantenous power without the presense of the ducted aircondition. It is clrealy shown the large reduction in the house electrical consumption.

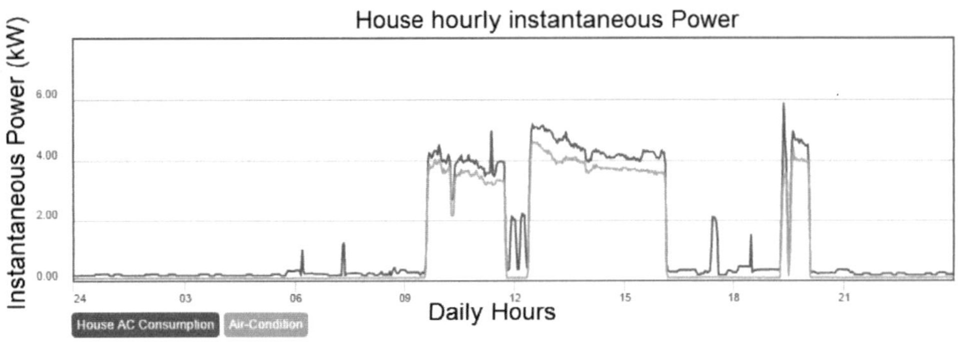

Fig. 7. House hourly instantaneous power with the ducted air-condition

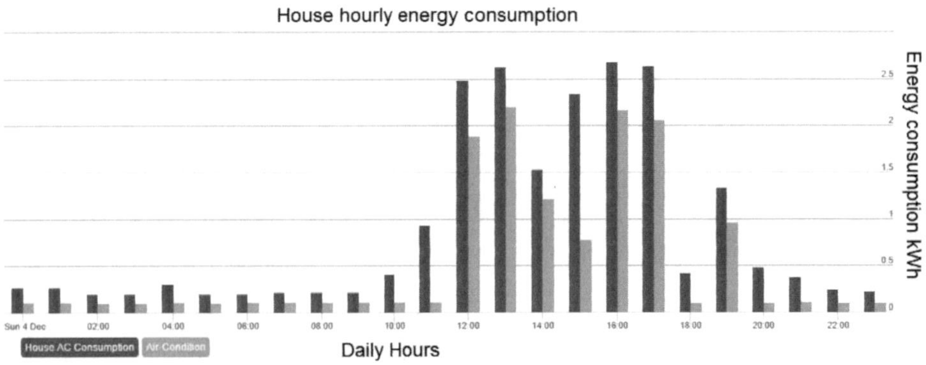

Fig. 8. House hourly energy consumption

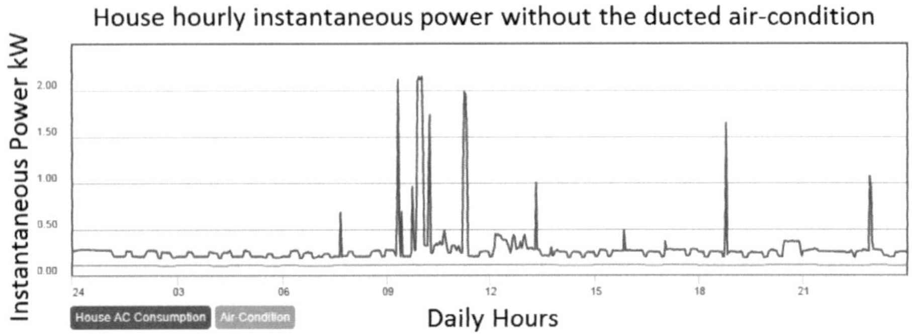

Fig. 9. House hourly instantaneous power without the ducted air-condition

V. CONCLUSION

The physical implementation of the system especially the roof design shows the large area that is made available for the installation of the PV system. Also, installing the panels at different direction assist with continuous power generation throughout the day. The installed system supports the theoretical study and is currently supporting the house since May 2016.

REFERENCES

1. K. Mitchell "Optimisation of the applications of sustainable energy systems" Ph.D. dissertation, School of Engineering and Industrial Design. Univ. Western Sydney. 2005.
2. K. Mitchell, M. Nagrial and J. Rizk "Simulation and optimization of renewable energy systems". Electrical Power and Energy Systems. Vol. 27. PP. 177-177. 2005
3. https://sustainabledevelopment.un.org/sdg7. Accessed on 2-12-2016 at 2:30PM
4. M.Nassereddine, J. Rizk, A. Hellany, M. Nagrial, A. Elrafhi, Z. Obeid and K. Hajar "Electrical Energy Management for adcance smart home systems: Introduction" 3rd International Conference on Renewable Energies for Developing Countries (REDEC), July, Lebanon 2016.

Performance Index: A Simple Selection Method of Appropriate Energy Storage for Specific Application

Hussein Ibrahim
InstitutTechnologique de Maintenance Industrielle (ITMI), Cégep de Sept-Îles, 175 rue de la Vérendrye Sept-Îles (Québec) G4R 5B7, Canada
hussein.ibrahim@itmi.ca

Mazen Ghandour, Sabine Saad
Faculty of engineering III, Lebanese University
Université Saint-Joseph,
Beirut, Lebanon
drmazengh@yahoo.com, sabinesaad@yahoo.com

***Abstract*— The efficiency and cost of renewable power systems using intermittent resources could significantly be improved by developing low cost, high efficiency and more sustainable energy storage systems. There are various types of storage methods, some of which are already in use, while others are still in development. A comparison study between energy storage options is presented in this paper. We have taken a look at the main characteristics of the different electricity storage techniques and their field of application (permanent or portable, long-or short-term storage, maximum power required, etc.). These characteristics will serve to make comparisons in order to determine the most appropriate technique for each type of application. Based the results obtained in this study, a "performance index" have been determined for each storage technology in order to facilitate the comparison between the different storage devices.**

***Keywords*— Energy storage; performance index; comparisons; batteries; compressed air; flywheel; hydrogen; constraints.**

I. INTRODUCTION

Currently, the electricity production is strongly centralized and, often, a long distance away from its end users. Load levelling is initially based on the prediction of seasonal and daily needs, but also, when energy production is not sufficient, on the contribution of secondary modes like thermal and hydraulic power plants. In fact, these plants also use stored energy: water for the pumped hydro storage plants, and fossil fuels for the thermal plants [1].

Decentralized electricity production and the fast growth of fluctuating and intermittent renewable sources (wind energy, solar energy, etc.) increase the difficulty to stabilize the electric grid, mainly due to an imbalance between supply and demand. It is therefore appropriate to generate the energy, transmit it, convert it, and then store it if need be. More than ever then, the energy storage has become a necessity. But it is very complex to store electricity as this requires costly and bulky equipment.

Decentralized electrical production from renewable energy has a major inconvenient: they fluctuate independently from demand. This means that they produce high energy generation when it is not needed and produce low energy generation when it is needed. Yet, the stability of a grid is based on the balance between supply and demand, where from the demand for further development of storage devices [2].

The fundamental idea of the energy storage is to transfer the surplus of energy produced by the power plants during the weak load periods to the peak periods (Fig. 1).

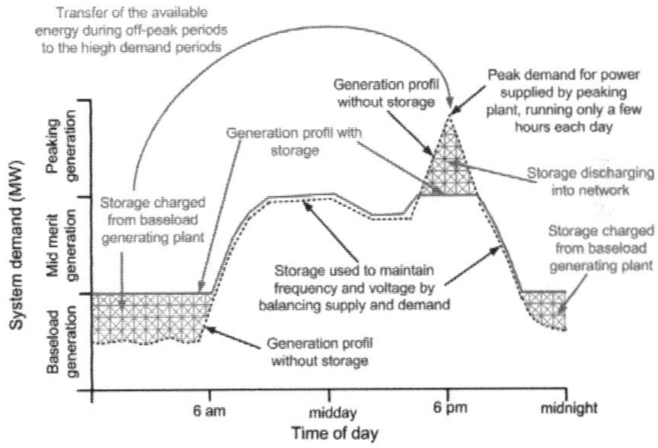

Figure 1: Principle of the energy storage

Initially, electricity must be converted into another form that can be stored (potential energy, mechanical, electrical, or chemical) and to be transformed back when needed. The stored energy should be rapidly converted on demand and used in a large variety of applications and load sizes. There exist different Energy Storage Systems (ESS) technologies; some of them are well studied and developed, while others are just emerging and waiting for new material technologies to make them cost-effective.

II. CLASSIFICATION OF ENERGY STORAGE SYSTEMS

There are many possible techniques for energy storage, using practically all forms of energy: mechanical, electrical, potential, chemical, and thermal. These have all been explored, leading to the birth of the techniques that will be cited herein. The energy storage technologies are many and answer to several technical and economic criteria, which vary considerably with the applications and type of needs. The comparison between these technologies is very difficult by the fact that their development levels vary greatly [3]. According to their end-uses, the energy storage systems (ESS) can be classified into four categories [4]:

1) *Low-power application in remote areas, essentially to supply transducers and emergency units;*
2) *Medium-power application in remote areas (individual electrical systems, town supply);*
3) *Grid connection application with peak sheaving;*
4) *Power-quality control applications.*

The first two categories are for small and medium scale systems used to supply isolated regions. In these categories, the energy could be stored as compressed air, chemical energy (battery), kinetic energy (flywheel), hydrogen (fuel cells), or in supercapacitors or superconductors. Categories three and four are for large-scale systems where the energy could be stored as chemical energy (accumulators, flow batteries), thermal energy (sensible, latent), gravitational energy (hydraulic systems), or compressed air.

III. Characteristics of Energy Storage Techniques

For a specific application, it is therefore essential to analyze the technical and economic characteristics of storage systems in order to identify the comparison criteria that will be used to determine the most appropriate energy storage technology.

The principal characteristics of ESS on which the selection criteria are based are the following:

A. Storage capacity

This is the amount of energy available in the storage system after charging. Because the discharge is often incomplete, the definition of storage capacity of ESS takes into account that the total energy stored is superior to that actually recovered. The usable energy, limited by the depth of discharge, represents the limit of discharge depth (minimum-charge state). In conditions of rapid charge or discharge, the efficiency deteriorates and the recovered energy can be much lower than storage capacity. On the other hand, self-discharge is the attenuating factor under very slow regime (Fig. 2).

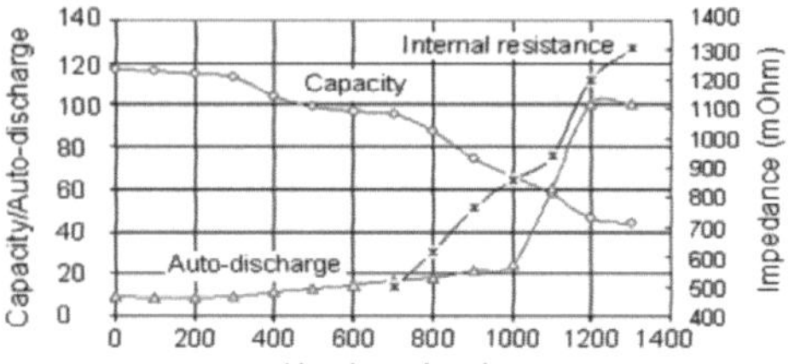

Figure 2: Technical characteristics of a nickel-metal-hydride battery [5]

B. Available power

This parameter determines the size of the electric component (motor/generator) in the stored energy conversion chain. It is generally expressed as an average value, as well as a peak value often used to represent maximum power of charge or discharge.

C. Power transmission rate

The power output or discharged power can be a limiting factor, determined by the power transmission rate. This delivery rate determines the time needed to extract the stored energy. The power must be available for delivery during peak hours. This means that the amount of energy which can be used, depends strongly on the system design, or on a fundamental limit of the storage apparatus [6].

D. Efficiency

This is the ratio between extracted energy and stored energy. This definition is often oversimplified because it is based on a single operation point [7]. The definition of efficiency must therefore be based on one or more realistic cycles for a specific application. Instantaneous power is a defining factor of efficiency.

E. Discharge time

This is the maximum-power discharge duration. It depends on the depth of discharge and operational conditions of the system, constant power or not.

F. Durability (cycling capacity)

This refers to the number of times the ESS can restitute, after each recharge, the energy rate it was designed. It is expressed as the maximum number of cycles N (one cycle corresponds to one charge and one discharge).

All ESS are subject to fatigue or wear by improper use. This is generally the main reason of aging, ahead of thermal degradation. The design of a ESS that considers the endurance of the unit in terms of cycles should be a primary importance when select a system. However, real fatigue processes are often complex and the cycling capacity is not always well defined. In all cases, it is strongly linked to the amplitude of the cycles (Fig. 3) and/or to the mean of the state of charge of ESS [8].

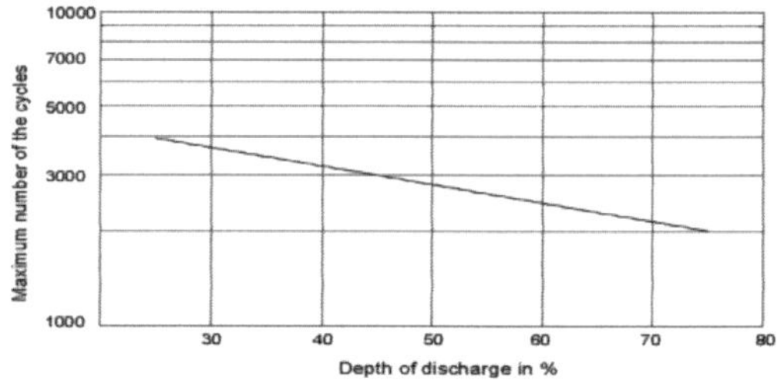

Figure 3: Variation of the cycling capacity with the depth of charge for a lead-acid battery [8]

G. Costs

Like any other investment, a storage system is an interesting project when total benefits surpass total expenses. The capital invested and costs of operation and maintenance are the most important factors to take into account for the entire lifecycle of the system.

Low-efficiency systems with low cycling capacity generally require a lowest initial investment. It is therefore important to take out an analysis of the estimated durability of the entire energy generation system including the ESS [9].

H. Autonomy

This refers to the maximum amount of time the system can continuously release energy. It is defined by the ratio between the energy storage capacity and the maximum discharge power. The autonomy of a system depends on the type of storage and the type of application.

I. Self-discharge

This is the part of the energy that was initially stored and which has dissipated over a given time and without any useful [10].

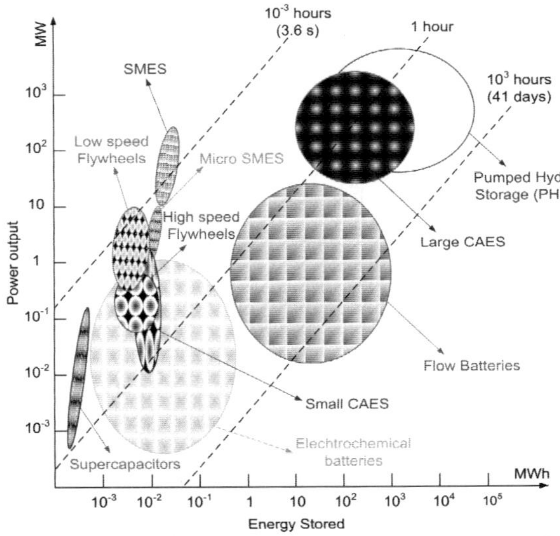

Figure 4: Repartition of the ESS according to energy needs and applications types [11]

J. Adaptability to the generating systems

To be strongly effective, an energy storage system needs to be closely adapted to the type of application (power generation in remote areas, grid connection, etc.) and to the type of production system (conventional, permanent, portable, renewable, etc.) that is meant to support (Fig. 4). In addition, the ESS must be harmonized with grid and with the entire electric system.

K. Energy density

This represents the maximum amount of energy that an ESS can be accumulating per unit of mass or volume. This parameter demonstrates the importance of the sizing of ESS (weight or volume) for certain applications especially in non-stationary applications, but less so for permanent applications.

L. Operational constraints

Especially related to safety (explosions, waste, bursting of a flywheel, etc.) or other operational conditions (temperature, pressure, etc.), they can influence the choice of a storage technology as a function of energy needs [8].

M. Reliability

Storage-system reliability is always an important factor because it is a guarantee of on-demand service [7].

N. Monitoring and control equipement

This equipment, on both the quality and safety of storage levels, has repercussions on the accessibility and availability of the stored energy.

O. Environmental aspect

While this parameter is not a criterion of the capacity of ESS, the environmental aspect of the product (recyclable materials) is a strong sales pitch. For example, in Nordic countries (Sweden, Norway), a important part of the population prefers to pay more for energy than to continue polluting the country [5].

P. Other characteristics

The ease of maintenance, simple design, operational flexibility (this is an important characteristic for the utility), fast response time for the release of stored energy, etc. [3].

IV. CASE STUDY: CHOICE OF AN APPROPRIATED STORAGE TECHNOLOGIE FOR WIND-DIESEL HYBRID SYSTEM

Canada is an immense country and while most of the population is concentrated near the US border, there is more than 300 000 peoples living in remote sites that are not connected to the main grid of electricity. In those isolated areas, electricity is generated using diesel gensets at very high economic and environmental costs. Since most of these regions have a good wind resource, an interested solution to improve the electrification of these sites is to couple wind turbines with the diesel power plants already in place. However, in the absence of a ESS, the penetration rate of wind energy must be kept low due to the constraints related to the operation of the diesel generator. This leads to dissipate an important part of the electricity produced from the wind turbine and reduces significantly the economic interest of the hybridization between diesel and wind energy. Wind energy penetration rate can be largely increased by the addition of an energy storage system. The utilization of the excess energy allows a much more stable power and the complete stop of the diesel generators when they are not needed and a close to optimal performance of it, which reduces their maintenance cost.

The required storage system should be easily adaptable to the wind-diesel hybrid system, available in real time and offer smooth power fluctuations. It is therefore necessary to analyze the fundamental characteristics (technical and economic) of storage systems in order to establish comparison criteria and select the best technology for the high penetration wind-diesel hybrid system.

A. Synthesis of the various energy storage techniques

To compare energy storage systems, Ragone's diagram is generally used to represent their technical and economic performances with the mass or volume energy densities [12]. This type comparison is conventional and it is particularly interesting for portable applications, for which mass is a critical aspect. However, for permanent applications, life expectancy and total costs (investment, operation and maintenance costs) are the most important criteria [2]. Thus, these comparisons, overly simplistic, do not reach a definitive conclusion about the validity of the choice of appropriated energy storage to associate with a wind-diesel hybrid system. For this reason, a thorough and detailed summary will be presented in the next sections explain this selection while based on techno-economic criteria listed above.

1) Electrochemical batteries energy storage (BES):

The lead batteries are the most economical investment at

the delivery time of the system. Unfortunately, these batteries are susceptible to misuse and have various disadvantages which need to conduct further research, especially to improve their life cycling (charging and discharging). It is in the best case, about of a thousand cycles (approximately 3 years), which is very insufficient in production systems whose lifetime is about 15 to 20 years. Moreover, the "sudden death" of lead batteries in end of life is hardly compatible with the requirements of reliable power in remote locations.

Alkaline batteries (Ni/Cd and Ni/MH) are much more efficient, robust but more costly. They are better adapted to low temperatures and used where high reliability is required. By cons, the Ni/Cd suffers, according to the method of use, from a "memory effect" which reduces the usable capacity. Technically, lithium batteries have the best performance. Indeed, self-discharge of Li/ion is low and their stored energy density is much higher. They also allow a cycling at high deep of discharge with less impact on performance compared to the nickel/cadmium and lead batteries especially. Primarily for cost reasons, these batteries are still quite uncommon.

Finally, when the batteries are not recycled, they are a major source of pollution. For all these reasons, the batteries will not be adopted for wind-diesel hybrid system (WDHS).

2) Hydrogen energy storage (HES):

It is impossible to use hydrogen produced from wind energy as a fuel for diesel engines because they are not equipped with ignition systems. By cons, the use of the stored hydrogen to produce the electricity from a fuel cell designed to operate in the absence of wind and replace diesel for certain periods, seems a good alternative. But the poor performance of the overall electrolyzer-fuel cell (35-40%), their unreliability, the prohibitive cost of capital and operating of this set and the safety aspect related to hydrogen storage, make sure they delay the acceptance of this technology by the market. For these reasons, the hydrogen storage will not be adopted for the WDHS.

3) Flywheel energy storage (FES):

This technology is not feasible to store large amounts of energy at long term. By cons, in the case of the supply of isolated sites, the flywheels can be used to store energy when production exceeds demand and improve the power quality where wind generation is close to the consumer level. Moreover, for a WDHS, the presence of an intermediary flywheel (by coupling the flywheel to diesel shaft) allows to reduce significantly the number of daily starts of diesel and increase the autonomy and the quality of the supplied current. Despite all these advantages, this technology has been excluded because it is a storage device for short time (a few tens of seconds to minutes) and requires, otherwise, massive materials to store energy, which increases the cost of system and of the used equipment. Moreover, this storage system will not allow a remarkable reduction in fuel consumption when the diesel functions in stand-by mode (periods of medium wind power penetration). May be added the issue of security related to the possibility of the collapse of the steering wheel.

4) Thermal energy storage (TES):

This type of storage, regardless of the technology used, can serve, using wind energy or excess heat from exhaust gases of diesel to produce hot water for heating or for use in the community where the WDHS exists. This system does not lead to reduce fuel consumption and GHG emissions. However, a better design may be offered and consists to couple the thermal energy storage (TES) to the wind-diesel-compressed air hybrid system. The TES will be used, in this case, as heat exchanger-storage device to recover the available heat in the air at the end of its compression and restore it afterwards during expansion of compressed air in the diesel engine or in an air motor. In conclusion, the TS will not be adopted as the main storage device associated with the WDHS but rather as an auxiliary system that improves the energy performance of the overall system.

5) Pumped hydraulic energy storage (PHS):

It is a large-scale storage and features a centralized production. So if it is available for remote sites, it will not be necessary to consider the use of diesel generators, especially if the wind potential is also available. By cons, storing small hydro requires abundant water resource in order to produce electricity; most of the criterion of geographical features is always present. For all these reasons, this technology has been excluded from the list of candidate technologies to be associated with the WDHS.

6) Redox batteries Energy storage (RBES):

Although the flow batteries can be used for small scale applications such as large scale, they are less mature and have yet to demonstrate their operation by hybridization with wind energy. Moreover, their prohibitive cost penalizes their integration into the market not to mention their contribution to the stopping of diesel is related to their autonomy, which also is depended to the reservoir volume of electrolytes in circulation and to the number of used cells (need large volumes and a high number of units). This increases the investment cost as well as the operating system. All these reasons exclude the possibility of adopting the flow batteries for WDHS.

7) Supercapacitors energy storage (SCES):

This type of storage is intended for small scale applications and its use at medium scale (case of WDHS) requires a series connection of several elements to achieve the required voltage. This will require more planning, more space and therefore more investment especially as the unit cost is quite high. The high self-discharge (5% per day), decreases the autonomy of the storage device and require a rapid consumption of wind energy stored before it is completely self-discharged without any use. This may slightly influence on the fuel consumption of diesel and its operating cost and make the overall system poorly operational and poorly harmonized. Consequently, this storage system will not be adopted for the WDHS.

8) Superconducting magnetic energy storage (SMES):

It is a storage technology for a very short duration and not feasible for medium-scale applications. Its complexity lies in the need for a refrigeration system, transformers and converters, and a large infrastructure. This greatly increases the cost and makes more complicate the operation of the system. Moreover, this type of storage is not too adaptable to WDHS and does not contribute to a remarkable reduction in fuel consumption. For these reasons, it is excluded from the list of the studied technologies.

After excluding all the technologies mentioned above, it is proposed to use the compressed air energy storage system (CAES) to associate with the WDHS. Indeed, the CAES is adaptable for both sources of power generation (wind / diesel). Moreover, the CAES is an interesting solution to the problem of strong stochastic fluctuations of the wind power because it offers a high efficiency conversion (60-70% for a complete charge-discharge cycle), uses conventional materials which are easy to recycle and is able to make an almost unlimited number of cycles [3], [13].

B. Validation of the choice of compressed air energy storage using the performance index method

1) Definition of the performance index

The analysis of the techno-economic criteria such as cost, energy density, lifetime, energy efficiency of each technology, etc.., allows evaluating a «performance index» and developing a yields diagram of different storage technologies. The performance index is the measure of the applicability of a storage technique to a specified application taking into account the characteristics presented in section 6 (for example, cost, efficiency, simplicity, life time, maturity, self-discharging, reliability, environmental impact, operation constraints, energy and power capacity, adaptability with wind-diesel system, contribution to reduce of fuel consumption, etc.) [14]. For another application than the power supply of a remote area, the values of the performance index can be different.

2) Method used to determine the performance index

The determination of performance index is done using a decision matrix that helps to balance by a factor varying from 5 to 10% the importance of each characteristic of the storage system with regard to the specific requirements of the envisaged application [15]. The development of this matrix takes into account the interaction between the selection criteria mentioned above, the reference data of storage technologies and the characteristics of WDHS and above the main objective: choice one technology that contributes to decrease the consumption fuel and GHG emissions of diesel generator.

In this study, more emphasis was given to cost, the contribution to reducing fuel and GHG emissions. A weight factor of 10% was then attributed for each of these criteria. By cons, the cyclability, autonomy, efficiency, simplicity, each has obtained a weighting equal to 7.5% for their medium importance [15]. Finally, because their lower importance, the following criteria: safety, self-discharge, storage, reliability, maturity, operational flexibility, control, and the ecological response times were weighted by coefficient of 5 % for each (Figure 5).

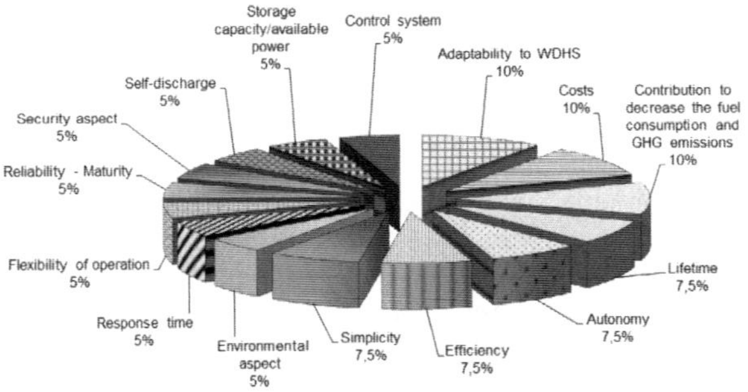

Figure 5: Allocation criteria in the decision matrix [15]

TABLE 1: ELEMENTARY DECISION MATRIX FOR THE CRITERION "ADAPTABILITY TO WDHS" [15]

Adaptability to WDHS	PHS	CAES	FES	BES	RBES	HES	TES	SCES	SMES	Total
PHS		0	0	0	0.5	0	0	0	0	**0.5**
CAES	1		1	1	1	1	1	1	1	**8**
FES	1	0		0.5	1	0.5	0.5	0.5	0.5	**4.5**
BES	1	0	0.5		1	0.5	0.5	0.5	0.5	**4.5**
RBES	0.5	0	0	0		0.5	0	0.5	0.5	**2**
HES	1	0	0.5	0.5	0.5		0	0.5	0.5	**3.5**
TES	1	0	0.5	0.5	1	1		1	1	**6**
SCES	1	0	0.5	0.5	0.5	0.5	0		0.5	**3.5**
SMES	1	0	0.5	0.5	0.5	0.5	0	0.5		**3.5**

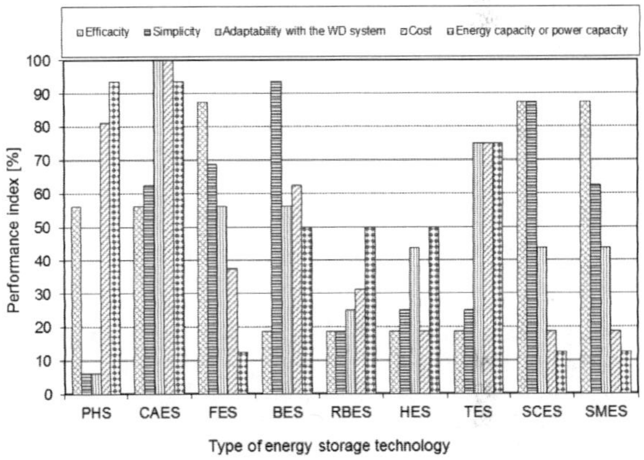

Figure 6: Comparison of the elementary performance index regarding to various criteria [15]

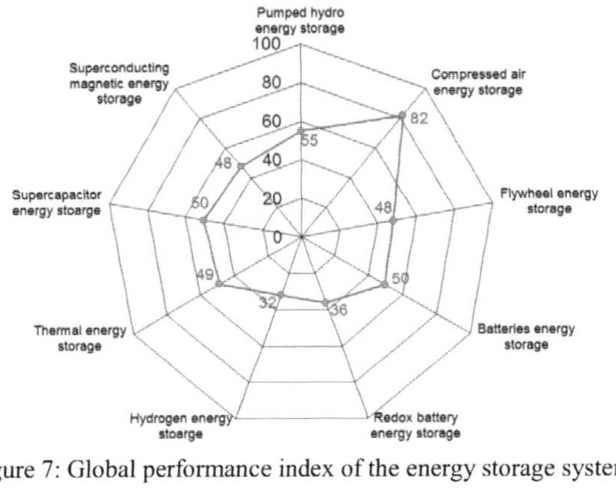

Figure 7: Global performance index of the energy storage systems

For each criterion, we attributed a value of either 0, 0.5 or 1 to each storage technology when comparing to another, respectively if the associated performance is worse (value of 0), equal (value of 0.5) or better (value of 1). In Table 1, we illustrate the comparison between the energy storage technologies with regard to the adaptability to the high penetration wind-diesel hybrid system. The analysis shows that flywheel is equivalent to batteries, supercapacitors, superconducting, thermal storage and hydrogen storage (coefficient of 0.5), while it is better (more adaptable to WDHS) than hydraulic storage and flow batteries (coefficient 1) [15-18]. On the other hand, the compressed air storage is more adaptable to SHED compared to the flywheel; the corresponding coefficient is then 0 for flywheel. The same method is applied to all the energy storage technologies for each of the 15 selection criteria, the results being 15 elementary decision matrixes corresponding to each criterion. The global decision matrix assembles the coefficients attributed to each energy storage system for each elementary decision matrix associated to each of the 15 selection criteria [15]. Considering the weight coefficients of every criterion in the calculation, we obtained the classification of performance of each used system. The elementary performance index (Fig. 6) and overall performance index (Fig. 7) can be obtained respectively from the elementary and overall matrix of decision by calculating the percentage of the total score obtained by each technology (global) and for each mentioned criteria (elementary) compared to the maximum total score (8) that a storage technology can be obtain. It is easy to establish from the Figure 7 that the CAES answers to the choice criteria with the better performance index, approximately 82 % [15-18].

V. Conclusion

A comparative analysis of different storage technologies currently in use was effectuated according to several criteria such as cost, energy density, specific power, contributing to reducing fuel consumption and GHG emissions, the lifetime and efficiency of each technology. This analysis was served to determine the performance index of each storage technology based on the nature of the project application. The determination of the performance index of each technology represents, despite its subjectivity resulting from the use of the decision matrix, a solution where we have some difficulty to choose a technology and where the constraint of time does not achieve a detailed modeling of the studied systems. This method, applied on wind –diesel hybrid system, showed that the CAES answers to the choice criteria with a performance index approximately 82 %. Other systems are also more or less effective but at the cost, simplicity, adaptability to the WDHS, the contribution to reducing fuel consumption and GHG emissions and duration of life that there is some difference. For these reasons, CAES technology was adopted to associate with the wind-diesel hybrid system.

References

[1] www.mines-energie.org/Conferences
[2] Hussein Ibrahim, Adrian Ilinca, Jean Perron, Investigation on the Performances of Energy Storage Techniques, Reference Number : CANCAM07-ETS-76, le 21ième congrès canadien de mécanique appliquée, Toronto, Canada, 3-7 juin 2007.
[3] H Ibrahim, A. Ilinca, J. Perron, Energy Storage Systems – Characteristics and Comparisons, Renewable & Sustainable Energy Reviews, May 2007.
[4] Couffin P, Perrin M. Le stockage des énergies intermittentes. Commissariat à l'Énergie Atomique (CEA). Clefs CEA (50–51), hiver 2004–2005. p. 136–8. /www.cea.fr.
[5] Faure F. Suspension magnétique pour volant d'inertie. Thèse de doctorat. Institut National Polytechnique de Grenoble,France; Juin 2003.
[6] Buchmann I, Choisir une batterie qui aura une longue durée, Cadex Electronics Inc. Canada, /http://www.buchmann.ca.
[7] Bonneville JM. Stockage cinétique. Institut de Recherche de l'Hydro-Québec, Varennes, Québec, Canada. Rapport interne; février 1975.
[8] Gergaud O. Modélisation énergétique et optimisation économique d'un système de production éolien et photovoltaïque couplé au réseau et associé à un accumulateur. Thèse de l'ENS de Cachan;décembre 2002.
[9] Anzano J.P., Jaud P, Madet D. Stockage de l'électricité dans le système de production électrique. Techniques de l'ingénieur, traité de Génie Électrique D4030; 1989.
[10] Messenger R, Ventre J. Photovoltaic systems engineering. Boca Raton, FL: CRC Press; 1999.
[11] H. Ibrahim, A. Ilinca, J. Perron, Systèmes de stockage de l'énergie pour les éoliennes, Colloque international «Énergie éolienne et sites éloignés», Îles-de-la-Madelein, Québec, Canada, 19-21 Octobre 2005.
[12] www.electricitystorage.org
[13] H. Ibrahim, A. Ilinca, J. Perron. Comparison and Analysis of Different Energy Storage Techniques Based on their Performance Index. IEEE-Electrical Power Conference 2007, Montreal, October 25-26.
[14] H. Ibrahim, A. Ilinca, J. Perron, Investigations des différentes alternatives renouvelables et hybrides pour l'électrification des sites isolés, rapport interne, UQAR, LREE–03, 2008.
[15] H. Ibrahim. Étude et conception d'un générateur hybride d'électricité de type éolien-diesel avec élément de stockage d'air comprimé. Thèse de doctorat, Université du Québec à Chicoutimi, Août 2010.
[16] H. Ibrahim, A. Ilinca J. Perron. Solutions actuelles pour une meilleure gestion et intégration de la ressource éolienne. CSME/SCGM Forum 2008 - The Canadian society for mechanical engineering, 5-8 juin, 2008, Ottawa, Canada.
[17] H. Ibrahim, A. Ilinca, M. Ghandour, M. Dimitrova, D. Boulay, C. Arbez, M. Adegnon. Investigation des diverses options de stockage d'énergie face à l'intégration des parcs éoliens dans les réseaux. 1ère Conférence Franco-Syrienne sur les énergies renouvelables, Damas, Syrie, Octobre 2010.
[18] H Ibrahim, R Beguenane, A Merabet. Technical and financial benefits of electrical energy storage. IEEE-Electrical Power and Energy Conference (EPEC), pp. 86-91, 10 october 2012.

Effect of pore geometry of mesoporous supports on catalytic performances in methane reforming

Joy Tannous, Leila Karam, Marie-Nour Kaydouh, Henri El Zakhem, Nissrine El Hassan*
Department of Chemical Engineering
University Of Balamand
Amioun, El Koura, Lebanon
nissrine.hassan@balamand.edu.lb

Sandra Casale
Laboratoire de Réactivité de Surface (LRS)
Université Pierre et Marie Curie, Paris 6
Paris, France

***Abstract*—Catalysts prepared using three dimensional SBA-16 silica support (composed of micropores and cage-like mesopores) were tested in the reaction of methane dry reforming, in comparison with 2D hexagonal mesoporous SBA-15 support. The samples were evaluated by N_2 sorption and X-Ray diffraction (XRD) for the assessment of their textural and structural properties. The reducibility was characterized by temperature programmed reduction (TPR). The catalytic performances were evaluated in methane dry reforming and spent catalysts (after reaction) were characterized for the evaluation of sintering and coke formation by TPH/MS, XRD and HR-TEM.**

***Keywords*—Mesoporous silica; methane dry reforming; pore geometry; sintering resistance**

I. Introduction

Natural gas can be converted to liquid hydrocarbons through two major steps; the first one is its conversion into syngas, which is then transformed by Fischer-Tropsch synthesis (FTS) into liquid hydrocarbons. The first step can be accomplished by steam reforming, partial oxidation, autothermal reforming or other reactions. An alternative way for this conversion is the dry reforming of methane DRM ($CH_{4(g)}+CO_{2(g)} \rightarrow 2CO_{(g)}+2H_{2(g)}$) that has additional advantages compared to the previously cited reactions: consumption of the greenhouse gas CO2 and production of a ratio H_2:CO close to one (good for increased selectivity in FTS).

DRM was performed over different types of catalysts especially noble metal catalysts like Pt, Ru, Rh, Pd and Ir that showed excellent activity and stability during reaction but these catalysts are expensive and rarely found. Nickel was as active as noble metals and thus became a promising alternative, especially that it is cheap and readily available [1-4]. However, Ni particles could still suffer from sintering or coke formation at high temperature, which drastically affects the catalytic performances. Consequently, researchers have intensified their efforts to solve these problems through several procedures such as incorporation of small amount of rare and alkaline earth metals over nickel based catalysts [5-7] or changing supports type and enhancing metal-support interaction [8,9].

Recent studies have shown that mesoporous SBA-15 silica support is promising for the confinement and stabilization of nickel in DRM reaction [10,11]. The addition of ceria to Ni/SBA-16 catalysts was studied by Zhang et al. [12] and the results showed that NiCe/SBA-16 was more stable, exhibiting higher conversion than Ni/SBA-16 in DRM. The studies in the literature were conducted under different experimental conditions (varying catalyst preparation and catalytic test conditions) and a direct comparison between SBA-15 and SBA-16 is not reported. Nevertheless, it is important to compare the two supports in order to check if the geometry of such materials can affect the performances of the resulting catalysts in the challenging DRM reaction.

II. Experimental

A. Catalysts preparation

Evaporation induced self-assembly [13,14] and sol-gel [15,16] methods are commonly used for the preparation of mesoporous materials. In this work, SBA-16(1) support was prepared using sol-gel technique following an established procedure of Cheng et al. [17]. In details, 5.66 g of the tri-block copolymer F-127 ($PEO_{106}PPO_{70}PEO_{106}$) (Sigma-Aldrich, P2443) were added to 250 ml of a 2M HCl. The mixture was stirred for 2 hours at 30°C. Then 19.81 g of TEOS (Sigma-Aldrich, 86578) were used as the silica precursor and the resultant mixture was stirred for 20 hours before filtration, washing with distilled water and drying in air at 110°C overnight. The support was next calcined at 500°C for 9 hours (at 2°C.min^{-1}).

SBA-16(2) support was prepared as reported by Dos Santos et al. [18]. 4 g of F-127 were added to 0.4M HCl. After complete dissolution, co-surfactant butanol (Sigma-Aldrich, 33065) was added and stirred at 45°C for 1 hour. This was followed by the addition of 19.3 g of TEOS and vigorous stirring at 45°C for another 24 hours. The obtained white precipitate was then filtered, dried in air at 100°C for 24 hours and then calcined at 550°C for 2 hours (at 2°C.min^{-1}).

Commercial SBA-15 mesostructured silica was used as the third support in this study (Sigma Aldrich, 777242).

Impregnation of nickel (5 wt%) on the three different supports was performed following "two solvents" loading method described elsewhere [19] and the calcination based on the optimized conditions reported in previous work [11]. The resulting samples were labeled as Ni/SBA-16(1), Ni/SBA-16(2), and Ni/SBA-15.

B. Catalysts characterization

The ASAP 2020 apparatus from Micromeritics was used to obtain the N_2-sorption isotherms at -196°C after proper degas under vacuum at 250°C for 2 hours.

X-Ray diffraction peaks and results were performed on a PANanalytical X'Pert[3] diffractometer using Cu Kα radiation (λ=1.5405 nm) at 2θ between 35 and 85°.

Temperature programed reduction (TPR) was tested using an AutoChem II, 2920 Micromeritics apparatus.

Temperature programmed hydrogenation (TPH) coupled to mass spectrometry measurements were done on spent catalysts while heating from 100 to around 900°C (at 10°C.min^{-1}) at atmospheric pressure under 5 vol% H2/Ar (30 ml.min^{-1}). Experiments were performed by connecting the AutoChem 2920 to an online quadrupole mass spectrometer (MKS, Cirrus 2). This technique is helpful in determining the type of deposited carbon in the spent catalyst according to its gasification temperature to CH4: Cα up to 350°C, 350 < Cβ <500°C and Cγ above 500°C [20].

A JEOL JEM-200 electron microscope operating at 200 keV (LaB6 gun) was used for Transmission electron microscopy (TEM) images.

C. Catalytic testing

Calcined catalysts were tested in a Hastelloy-X reactor (Microreactivity, PID Eng and Tech, internal diameter 9 mm). 200 mg of each catalyst were first reduced *in-situ* using 5 vol% H_2/Ar (30 ml.min^{-1}) at 650°C for 2 hours to ensure complete reduction of nickel oxide into Ni^0 metallic nickel. After reduction, the temperature was lowered to 200°C and then increased from 200°C to about 800°C (at 5°C.min^{-1}) again under an equimolar mixture of CH_4 and CO_2 using a gas hourly space velocity (GHSV) of 18 L.hr^{-1}.g^{-1}. The catalytic stability was measured by keeping the reactor at 650°C for 12 hours. Effluent gases from the reactor were analyzed online using an Inficon micro-GC equipped with a TCD and two types of columns (molecular sieves and plot-U) to detect H_2, CH_4, CO and CO_2 gases. The reactants conversions were calculated using the following equations:

$$CH_4 \text{ conversion } (\%) = 100 \times [CH_{4(in)} - CH_{4(out)}]/CH_{4(in)}$$

$$CO_2 \text{ conversion } (\%) = 100 \times [CO_{2(in)} - CO_{2(out)}]/CO_{2(in)}$$

III. RESULTS AND DISCUSSION

A. Structure and porosity of catalysts

N_2-sorption isotherms show a type IV isotherm for all catalysts, typical of mesoporous silica materials (data not shown). Ni/SBA-15 catalyst has H1-type hysteresis loop typical of 2-D hexagonal structure whereas both Ni/SBA-16(1) and Ni/SBA-16(2) show an H_2 hysteresis loop reflecting a cubic structure of the material with cage-like pores typical for SBA-16 parent [21].

TABLE I. BET SURFACE AREA AND CRYSTALLITE SIZE OF FRESH AND SPENT CATALYSTS

Catalyst	BET surface area from N_2 sorption isotherms ($m^2.g^{-1}$)	Crystallite size by XRD (nm)	
		Fresh catalyst (NiO)	*Spent catalyst (Ni^0)*
SBA-16(1)	408	-	-
Ni/SBA-16(1)	295	7.7	11
SBA-16(2)	336	-	-
Ni/SBA-16(2)	138	7.5	13.2
SBA-15	701	-	-
Ni/SBA-15	528	3.3	3.6

BET surface areas of the three different supports with their impregnated active phase (5wt %) are grouped in Table I. Concerning SBA-16 supports, their surface area was in the range of 300-410 $m^2.g^{-1}$ whereas SBA-15 support had much higher surface area (701 $m^2.g^{-1}$). For all SBA based samples, the surface area was decreased after nickel impregnation which is attributed to partially filled pores or to some collapse of the silica walls. But since the shape of isotherms for the impregnated SBA stayed similar to the parent SBA, a well-organized framework was preserved and it can be concluded that there was no destruction of support. Such destruction can take place under the preparation procedure used in this study, but the TEM images (Fig. 4(a)) prove that SBA-16 framework and pore structures remained intact, similar to the study done on DRM in [12].

XRD for calcined samples at wide angle (35 < 2θ < 85°) are represented in Fig.1A. Fresh Ni/SBA-15/16 catalysts exhibit three main peaks at 2θ = 37, 43 and 63° (identified by •) corresponding to NiO particles with face centered cubic (FCC) unit cell and reticular planes respectively indexed (111), (200), (220) (JCPDS 71-4750). Based on the Scherrer equation, the mean NiO particle sizes of each catalyst are found to be 7.7, 7.5 and 3.3 nm for Ni/SBA-16(1), Ni/SBA-16(2) and Ni/SBA-15, respectively (Table I). Since all SBA based catalysts are impregnated with the same amount of nickel (5 %wt), the difference in crystallite size means that nickel particles are better dispersed over SBA-15 support than on SBA-16.

B. Reducibility of Ni/SBA-16(1), Ni/SBA-16(2), and Ni/SBA-15

TPR profiles of reduced Ni/SBA catalysts (Fig. 2) show two hydrogen consumption peaks: a first peak between 300 and 450°C corresponding to free NiO particles easily reduced and a second peak attributable to Ni oxide species in strong interaction with the support and reduced at higher temperature, greater than 450°C [23]. With Ni/SBA-15, the main NiO species appear to be in enhanced interaction with the support. Whereas for Ni/SBA-16(1) and Ni/SBA-16(2), most of NiO where reduced at the lower temperature range.

https://doi.org/10.1051/matecconf/201817101003

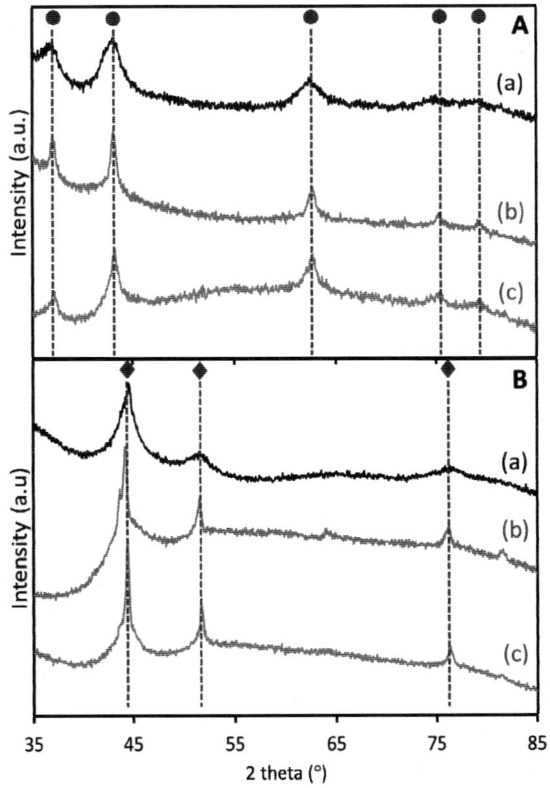

Fig. 1. X-ray diffraction patterns for (A) fresh and (B) spent catalysts (a) Ni/SBA-15, (b) Ni/SBA-16(1), (c) Ni/SBA-16(2)

This behavior could be due to the different availabilities of SiO_2 surface sites especially that Ni/SBA-15 has larger surface area than both Ni/SBA-16 catalysts and thus the smaller surfaces accessible on SBA-16 might limit the formation of mixed phases upon the addition of the nickel precursor [24].

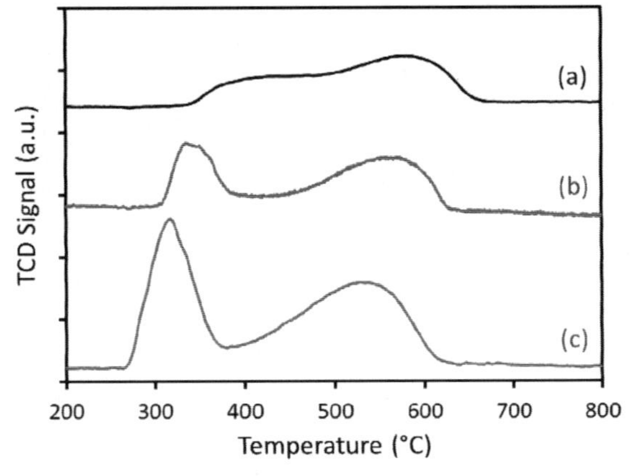

Fig. 2. TPR profiles for catalysts (a) Ni/SBA-15, (b) Ni/SBA-16(1), (c) Ni/SBA-16(2)

C. Activity and stability in DRM

Catalytic stability experiments (Fig. 3) were carried out at 650°C for 12 hours under stream. Ni/SBA-15 catalyst appears to be fully stable with preservation of CH_4 and CO_2 conversions (75 and 80% respectively). On the other hand, conversions on both Ni/SBA-16(1) and Ni/SBA-16(2) start at lower values than Ni/SBA-15 and are deactivation appears through time with an average conversion loss of about 20% and 30% respectively at the end of reaction. This reveals that SBA-15 structure could prevent deactivation and maintain catalyst stability, not to mention that CO_2 conversions remain higher than those of CH_4 due to the occurrence of RWGS. Reasons behind such behaviors resulting in deactivation will be discussed in the following section.

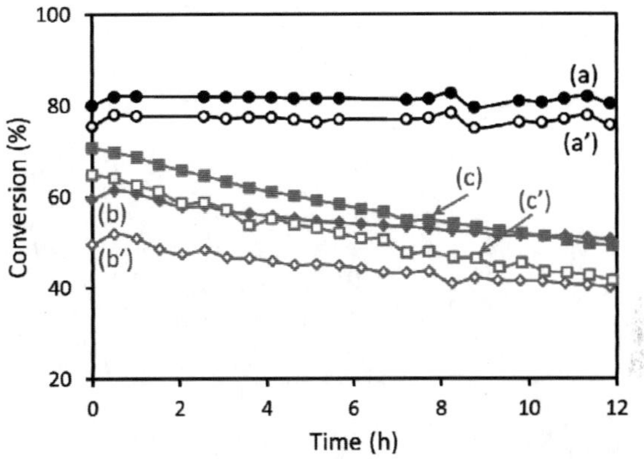

Fig. 3. Isothermal DRM experiments at 650 °C for (a-c) CH_4 conversions and (a'-c') CO_2 conversions: (a, a') Ni/SBA-15, (b- b') Ni/SBA-16(1), (c- c') Ni/SBA-16(2)

D. Physicochemical properties of the spent catalysts

To understand the reasons behind Ni/SBA-16 catalysts deactivation, XRD of spent catalysts, TEM and TPH/MS experiments were performed to check the presence of coke and/or sintering.

XRD for the 3 spent catalysts (Fig. 1(b)) show three peaks at 2θ = 44, 52 and 76° for planes (111), (200) and (220) respectively, corresponding to Ni metallic particles (identified by ♦, JCPDS 01-1258). This means that no re-oxidation of Ni^0 due to water formation through RWGS occurred during reaction [25,26] and therefore deactivation of Ni/SBA-16 is independent of this phenomenon. However, the peaks are much wider and almost invisible for Ni/SBA-15 catalyst than those for both Ni/SBA-16 catalysts that appear very intense. Table I shows that Ni particles over SBA-15 support remained small (3.6 nm) stressing on their high dispersion that is well retained even after test. Ni crystallite sizes for Ni/SBA-16(1) and Ni/SBA-16(2) are 11 and 13.2 nm respectively which highlights the occurrence of sintering, thus destabilizing and deactivating the catalysts as seen during catalytic test.

Fig. 4(b) of spent Ni/SBA-16(1) displays extremely large Ni nanoparticles with some carbon deposition. The weak interaction of the Ni species with SBA-16 could have facilitated their migration outside the mesopores and their aggregation at the surface at high temperature (compared to

 https://doi.org/10.1051/matecconf/201817101003

fresh ones, data not shown), thus reducing stability and catalytic performance [27]. However, on Ni/SBA-15 (Fig. 4(c)), Ni nanoparticles are still confined inside the pores with no migration to the external surface even after 12 hours under stream. In addition to this, Ni/SBA-15 shows high level of Ni dispersion that is restrained also after test which is in agreement with the small size of Ni particles (3.6 nm) obtained by XRD. Based on these findings and on TPR experiments, Ni species strongly interacting with the support can influence the catalytic stability of Ni/SBA-15. Therefore these catalysts were more stable and showed high resistance towards particles sintering, regardless of the coke formation that will be further elaborated in TPH experiments.

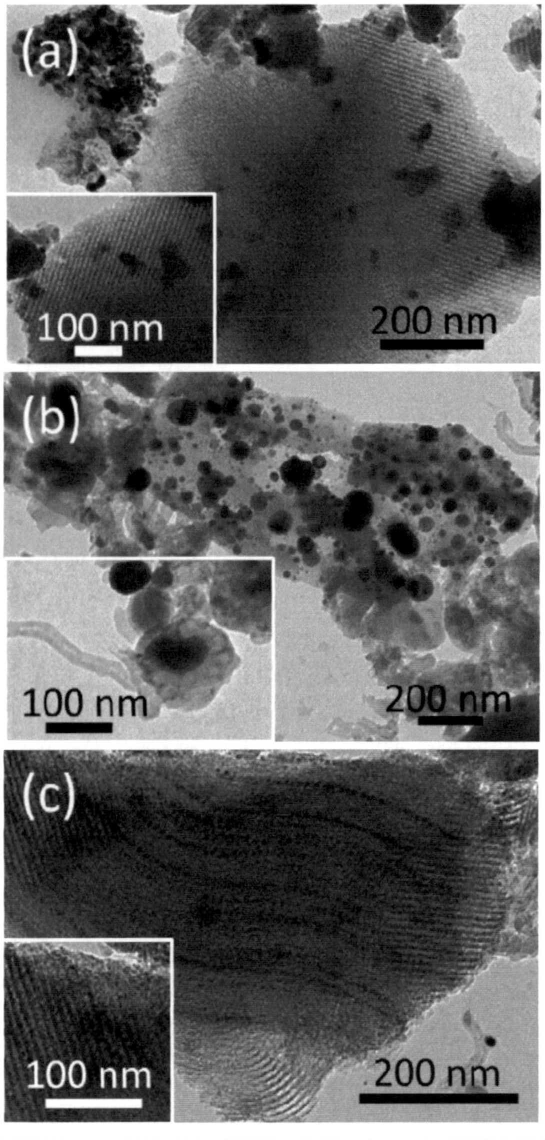

Fig. 4. TEM images of (a) calcined Ni/SBA-16(1) and spent (b) Ni/SBA-16(1), (c) Ni/SBA-15

In addition to metal sintering effect, coke formation might also have certain impact on catalysts behavior. TPH/MS experiments applied on the different spent catalysts determine the types of carbon deposits (Fig. 5). These experiments rely on the temperature at which methane (m/z = 16) is formed by hydrogenation of carbon species H_2 consumption.

From the three types of carbon discussed earlier, Cα is a reactive intermediate that does not have influence on catalyst stability; Cβ represents nanotubes and Cγ, a graphitic carbon, which is the most detrimental type responsible of catalyst deactivation [20,28,29]. Fig. 5 shows that Ni/SBA-15 catalyst (Fig. 5 (a)) contains the highest amount of carbon among others, but it is the only catalyst carrying the least amount of Cγ and the highest amount of Cα carbon. Cα is expected to be produced from thermal decomposition of methane and can lead after to the formation of CO by hydrogenation at low temperature [30]. Therefore, even if present at high concentrations, its effect is benign on catalyst performance and activity and this is emphasized by the high stability of Ni/SBA-15 during time on stream. Such behavior is also elaborated in a study done by Igor Luisetto et al. [31] where they show that Ni/CeO_2-Al_2O_3 prepared by co-precipitation method was highly active and stable in spite of the greater carbon deposition on its surface and related this behavior to the fact that Cα does not encapsulate Ni^0 sites which remain accessible to the reactants.

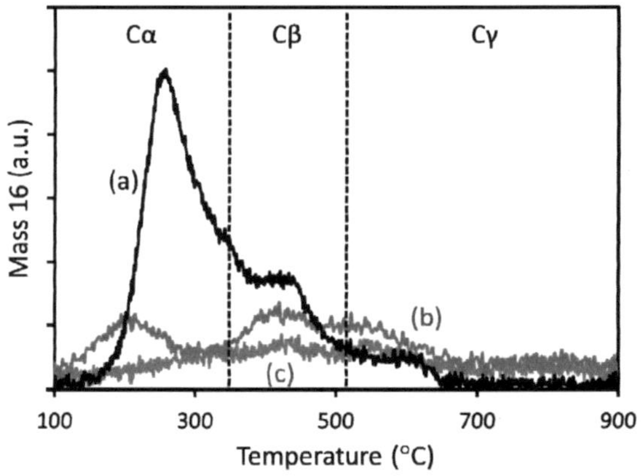

Fig. 5. TPH profiles for catalysts (a) Ni/SBA-15, (b) Ni/SBA-16(1), (c) Ni/SBA-16(2)

Thus, the main cause behind catalysts deactivation of Ni/SBA-16 is attributed to sintering, weak of metal-support interaction and low nickel dispersion as shown in [32]. It is important to mention that even if carbon formation does not alter catalyst performance, its mitigation and removal is a must because it can accumulate with time leading to reactor plugging and reaction perturbation.

IV. CONCLUSION

In this study, 5 wt% Ni were impregnated on commercial SBA-15 and synthesized SBA-16 supports and the properties of the resulting samples in addition to their catalytic performances were compared. While the SBA-15 2D hexagonal structure permits the stabilization of nickel inside the pores, SBA-16 cubic structure was not ideal for dispersion of nickel inside the pores nor for their stabilization under methane dry reforming. Ni/SBA-15 (without any promoter) are promising for DRM in opposition to Ni/SBA-16 catalysts where sintering is leading to rapid deactivation of the catalysts under stream.

ACKNOWLEDGMENT

Authors would like to acknowledge the help of Mrs. Hala Fallah in the XRD experiments.

REFERENCES

[1] M. F. Mark, W. F. Maier, "CO_2-reforming of methane on supported Rh and Ir catalysts. J. Catal., vol. 164, pp. 122-130, 1996.

[2] V. Y. Bychkov, O. V. Krylov, V. N. Korchak, "The Mechanistic Study of Methane Reforming with Carbon Dioxide on Ni/α-Al_2O_3",Kinet. Catal., vol. 43, pp. 86-94, 2012.

[3] J. L. Pinilla, I. Suelves, M. J. Lázaro, R. Moliner, J. M. Palacios, "Influence of nickel crystal domain size on the behaviour of Ni and NiCu catalysts for the methane decomposition reaction", Appl. Catal., A, vol. 363, pp. 199-207, 2009.

[4] M. Rezaei, S. M. Alavi, S. Sahebdelfar, Z. F. Yan, "Syngas production by methane reforming with carbon dioxide on noble metal catalysts", J. Nat. Gas Chem., vol. 15, pp. 327-334, 2006.

[5] J. Zhu, X. Peng, L. Yao, J. Shen, D. Tong, C. Hu, "The promoting effect of La, Mg, Co and Zn on the activity and stability of Ni/SiO_2 catalyst for CO_2 reforming of methane", Int. J. Hydrogen Energy, vol. 36, pp. 7094-7104, 2011.

[6] Z. Alipour, M. Rezaei, F. Meshkani, "Effect of alkaline earth promoters (MgO, CaO, and BaO) on the activity and coke formation of Ni catalysts supported on nanocrystalline Al_2O_3 in dry reforming of methane", J. Ind. Eng. Chem., vol. 20, pp. 2858-2863, 2014.

[7] S. Yasyerli, S. Filizgok, H. Arbag, N. Yasyerli, G. Dogu, "Ru incorporated Ni–MCM-41 mesoporous catalysts for dry reforming of methane: Effects of Mg addition, feed composition and temperature", Int. J. Hydrogen Energy, vol. 36, pp. 4863-4874, 2011.

[8] D. Liu, X. Y. Quek, W. N. E. Cheo, R. Lau, A. Borgna, Y. Yang, "MCM-41 supported nickel-based bimetallic catalysts with superior stability during carbon dioxide reforming of methane: Effect of strong metal–support interaction", J. Catal., vol. 266, pp. 380-390, 2009.

[9] S. K. Talkhoncheh, M. Haghighi, "Syngas production via dry reforming of methane over Ni-based nanocatalyst over various supports of clinoptilolite, ceria and alumina", J. Nat. Gas Sci. Eng., vol. 23, pp. 16-25, 2015.

[10] M. N. Kaydouh, N. El Hassan, A. Davidson, S. Casale, H. El Zakhem, P. Massiani, "Highly active and stable Ni/SBA-15 catalysts prepared by a "two solvents" method for dry reforming of methane", Microporous Mesoporous Mater., vol. 220, pp. 99-109, 2016.

[11] L. Karam, S. Casale, H. El Zakhem, N. El Hassan, "Tuning the properties of nickel nanoparticles inside SBA-15 mesopores for enhanced stability in methane reforming", J. CO_2 Utilization, vol. 17, pp. 119-124, 2017.

[12] S. Zhang, S. Muratsugu, N. Ishiguro, M. Tada, M., "Ceria-doped Ni/SBA-16 catalysts for dry reforming of methane", ACS Catal., vol. 3, pp. 1855-1864, 2013.

[13] C. Jeffrey Brinker, Y. Lu, A. Sellinger, H. Fan, "Evaporation-Induced Self-Assembly: Nanostructures Made Easy", Adv. Mater., vol. 11, No.7, pp. 579-585, 1999.

[14] D. Grosso, F. Cagnol, G. J. de A. A. Soler-Illia, E. L. Crepaldi, H. Amenitsch, A. Brunet-Bruneau, A. Bourgeois, C. Sanchez, "Fundamentals of Mesostructuring Through Evaporation-Induced Self-Assembly", Advanced Functional Materials, vol. 14(4), pp. 309-322, 2004.

[15] W. Hamd, S. Cobo, J. Fize, G. Baldinozzi, W. Schwartz, M. Reymermier, A. Pereira, M. Fontecave, V. Artero, C. Laberty-Robert, C. Sanchez, "Mesoporous α-Fe_2O_3 thin films synthesized via the sol-gel process for light-driven water oxidation", Phys. Chem. Chem. Phys., vol. 14, pp. 13224–13232, 2012.

[16] W. Hamd, M. Chavarot-Kerlidou, J. Fize, G. Muller, A. Leyris, M. Matheron, E. Courtin, M. Fontecave, C. Sanchez, V. Arterob, C. Laberty-Robert, "Dye-sensitized nanostructured crystalline mesoporous tin-doped indium oxide films with tunable thickness for photoelectrochemical applications", J. Mater. Chem. A, vol. 1, pp. 8217-8225, 2013.

[17] C. F. Cheng, Y. C. Lin, H. H. Cheng, Y. C. Chen, "The effect and model of silica concentrations on physical properties and particle sizes of three-dimensional SBA-16 nanoporous materials", Chem. Phys. Lett., vol. 382, pp. 496-501, 2003.

[18] S. M. L. dos Santos, K. A. B. Nogueira, M. de Souza Gama, J. D. F. Lima, I. J. da Silva Júnior, D. C. S. de Azevedo, "Synthesis and characterization of ordered mesoporous silica (SBA-15 and SBA-16) for adsorption of biomolecules", Microporous Mesoporous Mater., vol. 180, pp. 284-292, 2013.

[19] I. Lopes, N. El Hassan, H. Guerba, G. Wallez, A. Davidson, "Size-induced structural modifications affecting Co_3O_4 nanoparticles patterned in SBA-15 silicas", Chem. Mater., vol. 18, pp. 5826-5828, 2006.

[20] K. Jabbour, N. El Hassan, A. Davidson, P. Massiani, S. Casale, "Characterizations and performances of Ni/diatomite catalysts for dry reforming of methane", Chem. Eng. J., vol. 264, pp. 351-358, 2015.

[21] M. Thommes, K. Kaneko, A. V. Neimark, J. P. Olivier, F. Rodriguez-Reinoso, J. Rouquerol, K. S. Sing, "Physisorption of gases, with special reference to the evaluation of surface area and pore size distribution (IUPAC Technical Report)", Pure Appl. Chem., vol. 87, pp. 1051-1069, 2015.

[22] M. Kruk, C. M. Hui, "Thermally induced transition between open and closed spherical pores in ordered mesoporous silicas", J. Am. Chem. Soc., vol. 130, pp.1528–1529, 2008.

[23] T. Xie, L. Shi, J. Zhang, D. Zhang, "Immobilizing Ni nanoparticles to mesoporous silica with size and location control via a polyol-assisted route for coking-and sintering-resistant dry reforming of methane", Chem. Commun., vol. 50, pp. 7250-7253, 2014.

[24] J. A. C. Dias, J. M. Assaf, "Influence of calcium content in Ni/CaO/γ-Al_2O_3 catalysts for CO_2-reforming of methane", Catal. Today, vol. 85, pp. 59-68, 2003.

[25] C. Zhang, W. Zhu, S. Li, G. Wu, X. Ma, X. Wang, J. Gong, "Sintering-resistant Ni-based reforming catalysts obtained via the nanoconfinement effect", Chem. Commun., vol. 49, pp. 9383-9385, 2013.

[26] B. Huang, X. Li, S. Ji, B. Lang, F. Habimana, C. Li, "Effect of MgO promoter on Ni-based SBA-15 catalysts for combined steam and carbon dioxide reforming of methane", J. Nat. Gas Chem., vol. 17, pp. 225-231, 2008.

[27] D. Kantorovich, L. Haviv, L. Vradman, M. V. Landau, "Behaviour of NiO and Ni^0 phases at high loadings, in SBA-15 and SBA-16 mesoporous silica matrices", Stud. Surf. Sci. Catal., vol. 156, pp. 147-154, 2005.

[28] S. Zhang, S. Muratsugu, N. Ishiguro, M. Tada, "Ceria-doped Ni/SBA-16 catalysts for dry reforming of methane", ACS Catal., vol. 3, pp. 1855-1864, 2013.

[29] Z. L. Zhang, X. E. Verykios, "Carbon dioxide reforming of methane to synthesis gas over supported Ni catalysts", Catal. Today, vol. 21, pp. 589-595, 1994.

[30] Y. G. Chen, K. Tomishige, K. Fujimoto, "Formation and characteristic properties of carbonaceous species on nickel-magnesia solid solution catalysts during CH_4 CO_2 reforming reaction", Appl. Catal., A, vol. 161, pp. L11-L17, 1997.

[31] I. Luisetto, S. Tuti, C. Battocchio, S.L. Mastro, A. Sodo, "Ni/CeO_2–Al_2O_3 catalysts for the dry reforming of methane: The effect of $CeAlO_3$ content and nickel crystallite size on catalytic activity and coke resistance", Appl. Catal., A, vol. 500, pp. 12-22, 2015.

[32] M. M. Nair, S. Kaliaguine, "Structured catalysts for dry reforming of methane", New J. Chem., vol. 40, pp. 4049-4060, 2016.

Feasibility Study of Hybrid Wind-Diesel-Battery Power Generating Systems: Parametric and Sensitivity Analysis

Hussein Ibrahim
InstitutTechnologique de Maintenance Industrielle (ITMI), Cégep de Sept-Îles, 175 rue de la Vérendrye Sept-Îles (Québec) G4R 5B7, Canada
hussein.ibrahim@itmi.ca

Mazen Ghandour, Georges El-Jamal
Faculty of engineering III, Lebanese University, Université Saint-Joseph,
Beirut, Lebanon
drmazengh@yahoo.com, gfjamal@yahoo.com

***Abstract*— The renewables energies are being used to reduce the environmental pollution, combat the climate change and burning of fossil fuels. For remote or decentralized areas, where grid connection is very complex, renewable energy generation system can be a reliable and optimized source of energy. Moreover, wind-diesel-solar hybrid system technology promises lots of opportunities in remote areas which are far from the main grid and are supplied by diesel gensets. This paper is based on the analysis of a hybrid energy system for optimization. The analysis of the hybrid system is realized in the HOMER software package. The HOMER software was utilized as the assessment tool with modeling performed with hourly data of wind speed, solar radiation and load. In this study, the remote village of Tuktoyaktuk situated in Northwest Territories of Canada has been taken for the discussion of the optimization analysis of a hybrid energy generation system.**

***Keywords*— Hybrid power systems; renewables energies; wind energy; solar energy; batteries; optimization; sensitivity analysis; HOMER.**

I. Introduction

Many peoples in the world live in isolated and rural areas not connected to the main utility grid. It is very complex their supply by the distribution grid because of the high cost of transport and the distribution of energy to this remote areas [1]. The electric provisioning of these sectors is done by diesel generator systems [2]. However, this option is becoming obsolete due to high operating cost, issues of fuel transportation and environmental concerns. With increasing of the penetration of renewable energy technologies into the power market, remote area hybrid energy systems (RAHES) are considered as interested solutions to supply remote and regional areas [3]. Usually, RAHES systems consist of one or more renewable energy technologies with conventional generation schemes such as diesel [4]. The integration of conventional generating technologies to a RAHES system is fundamental due to uncertainties related to renewable sources, low inertial support and less reactive power support from renewable power plants [5]. However, to enhance the efficiency and profitability of the system, the part of the conventional power plant (ie. diesel generator) should be kept at a minimum [4].

Depending on the availability of resources, an appropriated generation mix can be envisaged to form a RAHES system. In the past, the hybrid systems have been considered as preferred for remote systems like radio telecommunication, satellite earth stations, or at sites far away from a conventional power system [6-9]. Today, there is a trend to update the existing one source system (PV, wind or hydro) into hybrid system [10].

The choice of renewable energy (RE) resources for such a power system is entirely dependent on the availability of the resources such as sun and wind. Among all these RE options, wind is classified as the fastest growing renewable energy technology in the energy industry [4].

Wind-diesel hybrid systems (WDHS) significatly reduce the need for storage of fuel, cost of fuel consumption, and greenhouse gas emission as opposed to their conventional diesel only counterpart. With increasing introduction of wind turbines in WDHS, system stability is becoming a crucial issue to the power supplier. Indeed, the variable, stochastic and erratic characteristics of wind energy create several problems when a wind generating system operates in its standalone operation mode that may lead to frequency and voltage perturbations [11-12]. The most difficult issue is to assess the capacity adequacy of the hybrid system in addressing the electricity demand of the consumers. However, the well-developed techniques applied to conventional generation system reliability evaluation, associate fixed capacity outputs to generating units and cannot be readily extended to include wind energy sources that have highly fluctuating capacity levels.

Therefore, it is necessary to integrate an energy storage system (ESS) to improve the performance of such a power system [11]. The power mismatch between generation and demand can be delivered or stored using an ESS system.

When selecting an ESS system for a remote hybrid energy system, the ESS technology, efficiency, reliability, cost, autonomy, and rating of the ESS are the most important factors to be considered. Among the different energy storage technologies available, battery can be considered as the most appropriate energy storage option for an autonomous power system due to its high energy density capacity [13]. A dump load also plays an important role in a RAHES system. In a real

RAHES system, a dump load can be a space heater or a water heater that absorbs the excess of energy which cannot be utilized by an ESS.

II. Remote Area Hybrid Energy System Description

The RAHES system considered in this paper which consists of a wind power plant, main loads and dump load, battery storage system, diesel power plant, and its controller, is shown in Figure 1. It is important to mention that the control algorithm is not presented in this paper because the main objective of this work is to conduct a sensitivity and parametric study of the RAHES. However, we assume that the active power of the system has been controlled with objective to minimize the imbalance of the RAHES's active power while extracting maximum power from wind turbines.

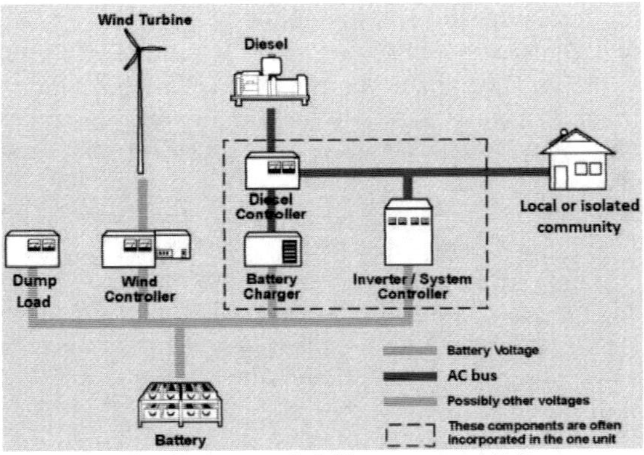

Figure 1: Schematic of remote area hybrid energy system [14].

In the present work, HOMER Micropower Optimization Model was used to conduct the designing, sensitivity and parametric study and techno-economic analysis of wind-diesel-battery hybrid systems. HOMER is general purpose hybrid system design software that facilitates design of power systems for stand-alone applications.

III. HOMER Micropower Optimization Model

The Hybrid Optimization Model Energy Resource, otherwise known as HOMER, is a computer model tailored for the simulation of hybrid power systems. Originally designed at the National Renewable Energy Laboratory (NREL) for the village power program, HOMER is now licensed to HOMER Energy [15]. HOMER simplifies the task of evaluating designs of both grid-connected and off-grid power systems for several applications. HOMER is a software that designs the most optimized and cost effective hybrid energy system component configuration after a long series of hourly simulations considering some parameters and component prices. The wind speed data, solar radiation data, load profile, prices of the system components have to be provided to the software.

HOMER achieves hundreds of hourly simulations to ensure the best possible matching between supply and conception in order to design the optimum system [16]. To analyse the effect of modifications of some parameters such as, wind speed variation, solar radiation, PV investment cost variation and diesel fuel price variation on the optimum result, sensitivity analysis can also be done. HOMER simulates the operation of a system by making energy balance calculations for each of the 8,760 hours in a year.

For each hour, HOMER compares the total demand (thermal and electric) in the hour to the energy that the system can supply in that hour, and calculates the energy flows to and from each component of the system [17].

For systems that include conventional generators or batteries, HOMER also decides for each hour how to operate the generators and whether to charge or discharge the batteries. After simulating all of the possible configurations of the system, HOMER displays a list of configurations, classified by lifecycle cost (LCC) that can be used to compare the different design options of the system [17].

Simulation, Optimization and Sensitivity analysis are the three major actions run by HOMER. In the simulation process, different power system configurations for every hour of the year are generated with their technical feasibility and LCC [18]. In the optimization process, HOMER selects one system configuration out of all configurations generated in the simulation process that satisfies all technical constraints and has the lowest LCC. In the sensitivity analysis, multiple optimizations are achieved on the selected configurations by Homer with a range of parameters that are supposed uncertain and there can be affect the model inputs with time. For the different variables known to the system designer- that is, the mix of system components and their respective quantity and size - the optimization process allows to calculate the optimal value. There are, however, also unknown factors such as uncertainties or changes in the variables outside the designer's control (for example, the average wind speed, rises in the fuel price or). The effects of these can be analysed with the help of the sensitivity analysis [18].

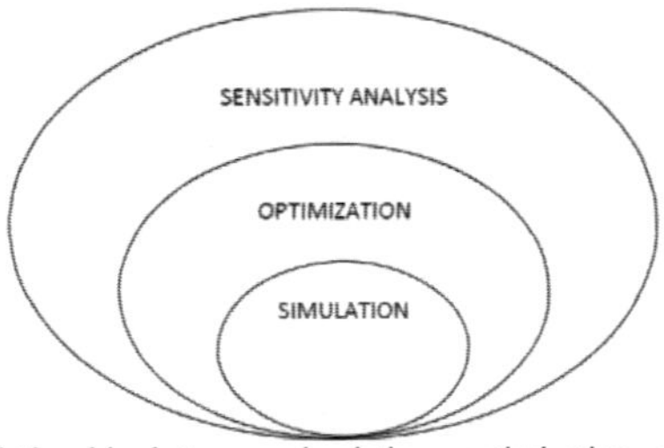

Figure 2: Relationship between simulation, optimization, and sensitivity analysis in HOMER [19].

Figure 2 shows how simulation, optimization and sensitivity analysis are interrelated. The oval figure enclosed in another oval figure shows that one optimization consists of multiple simulation and one sensitivity analysis consists of multiple optimizations.

IV. Physical Modeling And System Inputs

This section provides details about the physical operation that HOMER models for its simulation process.

The subsections describe the diverse input parameters HOMER requires to model the system: the chosen energy components to produce electricity, the energy demand (village load) that the system has to serve, the various energy resources associated to the selected components, and how this hybrid configuration operates to supply the loads.

A. Electrical load

In this study, the data of electrical load are taken for the village of Tuktoyaktuk situated on the Arctic Ocean in the extreme north of Canada's Northwest Territories. The average electric load of Tuktoyaktuk is 506 kW while its maximum load is about 851 kW.

B. Wind Resource

Accurate wind speed data is primordial to the precision of any wind-diesel hybrid system model. The data used in the HOMER simulations was registrated between January and December 2007. The characteristics of the wind speed throughout the year can be observed from the figures 3 and 4. Figure 3 shows the variation of the monthly wind speed in the site (Tuktoyatuk village). In addition, the wind speed variation over a period at the site is described by the probability density function (Weibull distribution) shown in Figure 4.

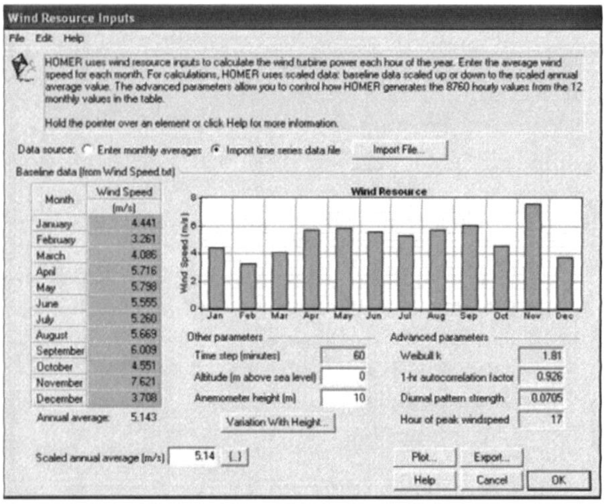

Figure 3: Wind speed data throughout the year.

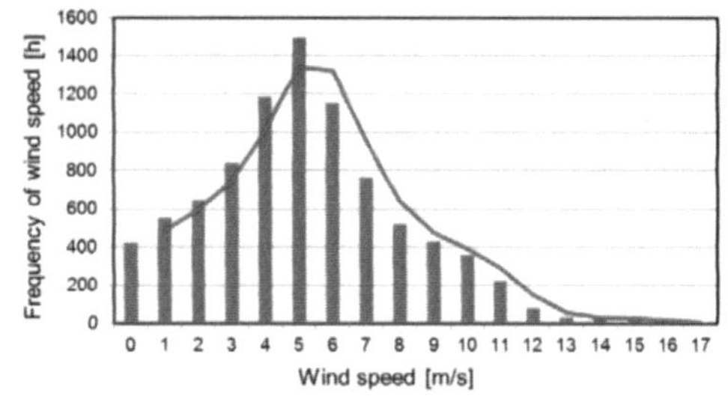

Figure 4: Wind speed distribution throughout the year

The wind resource during this period has an overall mean wind speed of 5.14 meters per second. The highest mean wind speed over an hourly span was 7.62 meters per second. The previous figure shows the wind speed distribution at Tuktoyaktuk village. This distribution is generated from the wind resource data input and can be characterised by four factors:

- *The Weibull shape factor (K):* This factor measures the distribution of wind speed over a year. The value of K is taken 1.81 in this study.
- *The autocorrelation factor:* This factor measures the randomness of the wind. Higher values indicate that the wind speed in 1 hour tends to depend strongly on the wind speed in the previous hour [16]. Lower values mean that the wind speed tends to fluctuate in a more random fashion from hour to hour. The autocorrelation factor value is taken as 0.926.
- *The diurnal pattern strength:* This factor measures the wind speed variation over a day. It shows strongly the wind speed depends on the time of day [16]. Higher values indicate that there is a strong dependence on the time of the day. In this study, 0.07 is used.
- *The hour of peak wind speed:* It is simply the time of day tends to be windiest on an average throughout the year. In this study, 17 is used as the hour of peak wind speed.

Figure 3 shows that there are 17 hours of peak wind speed. The wind speed variation over a day is 0.25 and the randomness in wind speed (autocorrelation factor) is 0.85.

Figure 4 shows that the most probable wind speed range 2–8 m/s, occurs approximately 70% of time. However, the high value of wind speed above this occurs only for limited period of time. This shows the site to be appropriate for wind power production, especially with application of low speed wind turbine.

C. Wind Turbine

HOMER has the data for a various manufacturers of wind turbines and their sizes already built into the program. For this simulation, several types of wind turbines have been studied such as: Enercon-E33, Fuhrländer100, Fuhrländer250, NorthenPower100 and WES 30 (Figure 5).

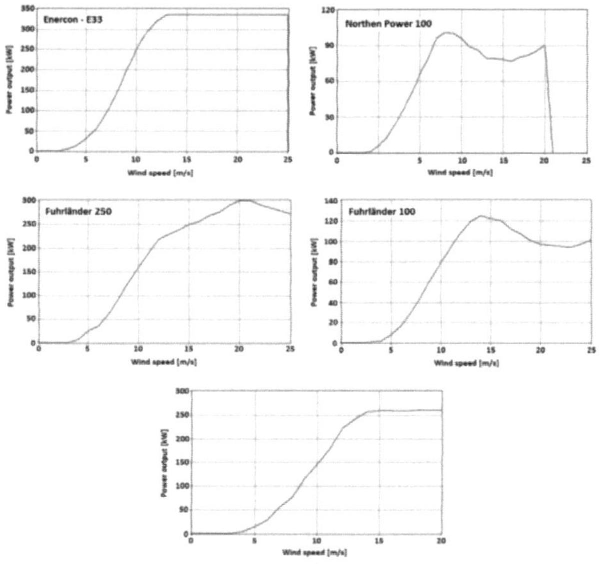

Figure 5: Power curves of five technologies of wind turbines

Enercon-E33 wind turbine was chosen for the size of its rated output (330 kW). A wind turbine of this size is typical for a wind-diesel-battery hybrid system with a 851 kW peak load. Moreover, the power curve of the turbine is very specific because the power generated increases rapidly with the wind speeds and reaches the rated power at a speed of 12m/s and stays stable at very high wind speed compared to other turbines. Thus, Enercon-E33 is characterized by a high power at low wind speeds which makes it advantageous for the case of Tuktoyaktuk.

TABLE 1: RESULTS OF SIMULATION RESULTS OF THE CASE OF WIND-DIESEL WITHOUT ENERGY STORAGE

Wind turbine	Rated power (kW)	Quantity	Wind energy production (kWh)	Fraction of wind energy (%)	Energy cost ($/kWh)	Excess energy (kWh)
E-33	330	3	1,840,020	15%	0.578	1,173,411
FL 100	100	8	1,552,419	13%	0.592	975,926
FL 250	250	3	1,208,774	12%	0.586	695,677
NP 100	100	8	1,243,942	11%	0.6	748,983
WES 30	250	3	1,074,283	10%	0.59	613,971

The results of the techno-economic analysis conducted with HOMER on a wind-diesel hybrid system without storage are presented in the Table 1. As we can see, the obtained results confirm the profitability of the choice of the E-33 wind turbine for Tuktoyaktuk village (minimum cost per kilowatt-hour: 0.578 $/kWh).

D. Diesel Generator

Generally, the ratings of the diesel generator are determined by the load but, currently, three diesel generators are used to supply the village by electricity. Their rated power output is 725, 960 and 1.400 kW despite the load varies between 240 and 850 kW. Preliminary simulation results of the current state in Tuktoyaktuk (Figure 6) show that the most used scenario is to operate the first two diesel generators where the third (1400 kW) rarely works.

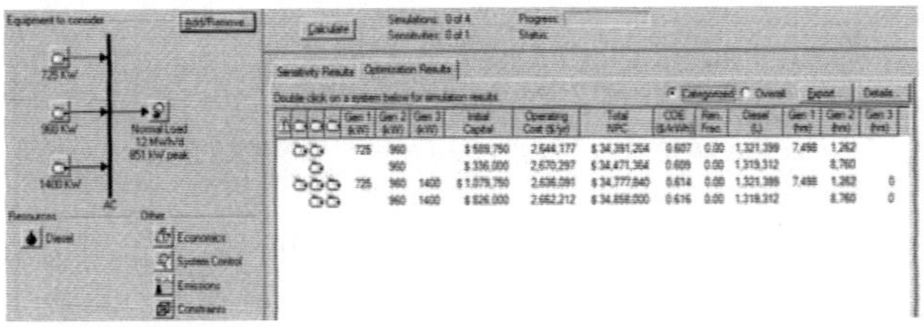

Figure 6: HOMER simulation results of the reference case with different configurations of diesel generators

Other configurations of diesel generators have been proposed and simulated on HOMER in order to optimize the current context of the electricity production in Tuktoyaktuk. HOMER has the data for generic diesel generators built in. However, the results of this analysis, summarized in Table 2, show that the configurations are ineffective and not very attractive without the contribution of renewable sources. Indeed, the best cost of kilowatt-hour obtained with these configurations is $ 0.694 when two diesel generators of 580 kW each are used, and the high cost is 0.846$/kWh using three generators of 400 kW each. For this reason, we have concluded that it is simpler and more efficient to install a single diesel generator supplying the village. This generator's main role is to provide electricity to the village in case of low wind energy penetration rate and during peak periods.

TABLE 2: RESULTS OF SIMULATION RELATED TO THE DIESEL PLANT CONFIGURATION

Configuration	Cost of kWh ($/kWh)	Fuel consumption (Litres/year)	CO_2 (Kg/an)
1x725kW & 1x960kW	0.607	1,321,399	3,479,678
1x960kW	0.609	1,319,312	3,439,633
2x580kW	0.690	1,285,766	3,385,844
3x400kW	0.846	1,567,474	4,127,676

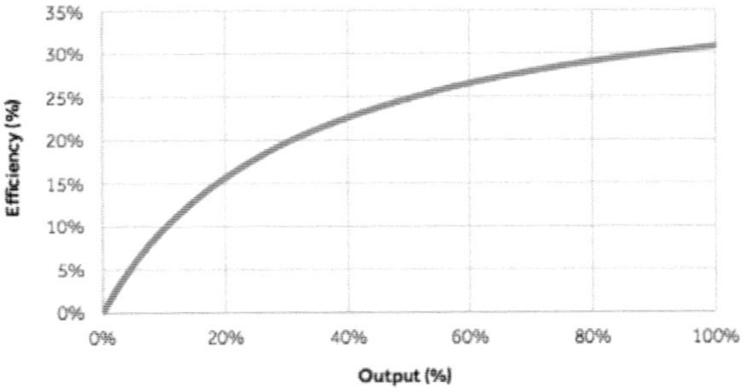

Figure 7: Diesel generator efficiency curve.

Taking into account the operation and maintenance purposes, a diesel generator of 1200 kW rated power (RP) is used. In this case, the size of the diesel generator has been selected such that it will meet the minimum loading condition in order to avoid low load factor operation and will supply the maximum load (851 kW) which represents 70% of the RP of DG. Therefore, 30% of diesel power are available such as spinning reserve and also can be used to charge the energy storage system (if available). A second generator of the same power must be provided to be used only in the emergency case or if the first generator is out of service. The diesel gensets are supposed to have an operating lifetime of 15,000 h with a minimum load rate of 30%. Figure 7 shows the efficiency curve relating the fuel consumption to the power output for the diesel gensets. This curve demonstrates two aspects of diesel power plant. Firstly, in general, the efficiency of diesel genset is very low. Indeed, operating at a peak efficiency of around 32%, 68% of the energy supplied to a diesel generator is then wasted in heat. Secondly, there is a pronounced decreasing in efficiency with a reduction in power output. For this reason, a constraint is placed on the system within HOMER to limit the minimal power output of the diesel gensets to 30% of their overall capacity.

The fuel consumption is modeled in HOMER by a linear curve characterized by a slope and intercept at no load. For a capacity range of 300 kW (minimal limit) to 1200 kW, the slope and the intercept are 0.215 l/h/kW and 0.051 l/h/kW respectively.

E. Battery Energy Storage System

Battery is a storage unit fundamental for storing electrical energy for maximum utilization of intermittent renewable energies [20]. In addition, the main purpose of having a battery energy storage system is to minimise the demand-generation mismatch in the absence of the diesel power plant. The lead-acid battery, often used in RAHES, is the batteries technology chosen for this study. Modeling of lead-acid batteries for real time analysis of RAHES must taking into account the folling parameters: (a) rate of charge/discharge, (b) ambient temperature, (c) battery storage capacity, (d) state of charge, (e) life and other internal phenomenon, such as gassing, double layer effect, self-discharge, heating loss, and diffusion [20].

A battery bank, composed from several identical batteries connected in series and/or parallel, must be sized correctly in order to obtain the desired capacity taking into account the load and the excess energy provided by wind power system.

Considering the profile of the energy consumption and the power curve generated by the wind, it was deduced that the maximum difference between these two profiles is about 696kW. The preliminary design must provide the installation of a battery bank able to absorb this maximum power.

The type of the batteries used in this study is of 'Trojan L16P ". The technical characteristics of this battery are presented in the figures 8 and 9. It is important to mention that we will need about 6,500 batteries to be able to store completely the maximum excess power because each battery of this type have 6V as voltage, 18A as current and 360Ah as individual storage capacity.

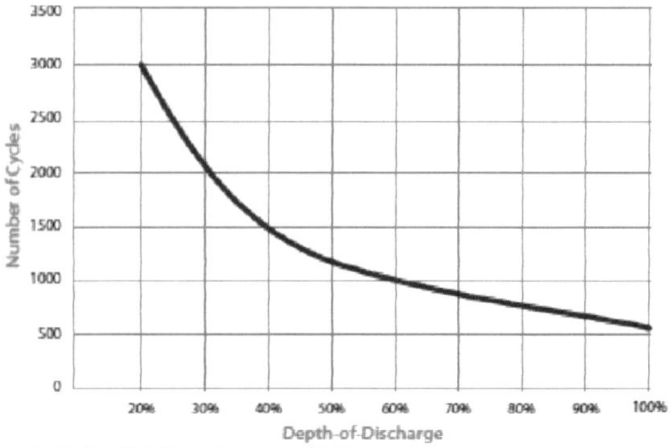

Figure 8: Trojan L16P performance [21]

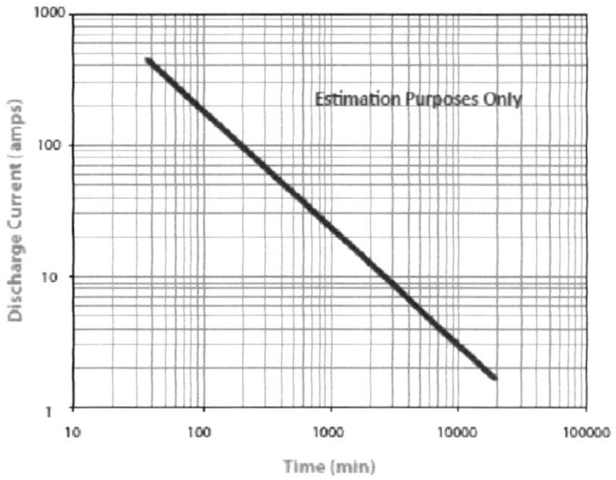

Figure 9: Typical cycle life in a stationary application [21]

The final outcome of preliminary selection of the components of the wind-diesel-battery hybrid system (WDBHS) is presented in the Table 3. The simulation results of this configuration are presented in the Table 4.

TABLE 3: COMPONENTS OF THE PRELIMINARY CONFIGURATION

Equipment	Capacity (kW)	Quantity
Enercon E-33	330	3
Diesel generator	1,200	2
Power converter	700	1
Batteries – Trojan L16P	0.108	6,500

TABLE 4: RESULTS OF THE PRELIMINARY CONFIGURATION OF THE WDBHS

Description	Value
Investment ($)	$3,821,500
Price of kWh ($/kWh)	0.535
Fraction of renewable energy (RE) (%)	22%
Fuel consumption (Liters)	1,024,371
Operation time of the 1st diesel generator (hr)	3,046
Operation time of the 2nd diesel generator (hr)	0
Excess Energy (kWh)	96,784
Emissions (Kg CO_2/year)	2,697,507

These results show that the excess energy dissipated without useful utilisation has largely decreased (by approximately 91.75%) compared to the scenario of wind-diesel without storage, which certainly increases the profitability of the project. However, it is possible to further optimize the WDBHS performance by acting on the configuration of the system. The optimization scenario will be presented in the followed sections.

V. OPTIMIZATION OF WDBHS

A. First optimization scenario

A preliminary optimization was conducted by modifying the important variables such as the number of wind turbines and the number of batteries in order to determine the best configuration based on the cost of the produced kWh. The obtained results are shown in the Tables 5 and 6. It is to note that the number of diesel generators has not changed compared to the primary configuration.

TABLE 5: COMPONENTS OF THE BEST CONFIGURATION BASED ON THE FIRST OPTIMIZATION SCENARIO

Equipment	Power (kW)	Quantity
Enercon E-33	330	10
Diesel generator	1,200	2
Power convertor	700	1
Batteries – Trojan L16P	0.108	8,000

TABLE 6: SIMULATION RESULTS OF THE BEST CONFIGURATION OF THE WDBHS BASED ON THE FIRST OPTIMIZATION SCENARIO

Description	Value
Investment ($)	$7,095,000
Price of kWh ($/kWh)	0.424
Fraction of renewable energy (RE) (%)	60%
Fuel consumption (Liters)	529,667
Operation time of the 1st diesel generator (hr)	1,606
Operation time of the 2nd diesel generator (hr)	0
Excess Energy (kWh)	2,903,407
Emissions (Kg CO_2/year)	1,394,787

From the results presented in the Tables 5 and 6, a significant reduction, about 20%, can be observed on the cost of produced kWh which decreases from 0.535$/kWh to

0.424$/kWh. Also, a radical rise in the fraction of renewable energy (i.e. wind energy penetration rate) and in the excess energy (i.e. dissipated energy) was also noted. This is due to the large number of turbines used in this scenario. The storage capacity could then increase to be able to absorb the all excess energy, but it would increase the investment cost, and therefore the cost of produced energy ($/kWh).

Moreover, by analysing the costs of the system, an increasing about 85% of the investment cost, equivalent to 3,273,500$, can be observed compared to the preliminary selection of components of the hybrid system. Hence the need for an intermediate solution taking into account the impact of the investment part in the total cost of the system should be investigated.

B. Second optimization scenario

The development of this solution needs to take into account the cost per kWh and the investment cost. These two factors are the key element with other secondary factors, in determining the profitability of wind-diesel-battery hybrid system (WDBHS) [22-25].

We conducted tens simulations of WDBHS with a number of wind turbines varying between 2 and 10. The purpose was to observe variations in investment costs and of the kWh, and the saving achieved compared to the baseline scenario. The obtained results are shown in Table 7 and figures 10 and 11.

TABLE 7: SIMULATION RESULTS OF THE BEST CONFIGURATION OF THE WDBHS BASED ON THE FIRST OPTIMIZATION SCENARIO

Number of used wind turbines	Investment ($)	Fuel consumption (L)	$/kWh	Saving (%)
2	$3,011,000	1,179,446	0.584	4.3%
3	$3,596,500	1,027,839	0.532	12.8%
4	$4,182,000	906,314	0.496	18.7%
5	$4,617,500	816,492	0.471	22.8%
6	$5,203,000	726,736	0.452	25.9%
7	$5,488,500	686,448	0.443	27.4%
8	$6,524,000	589,377	0.433	29.0%
9	$6,809,500	549,926	0.425	30.3%
10	$7,095,000	529,667	0.424	30.5%

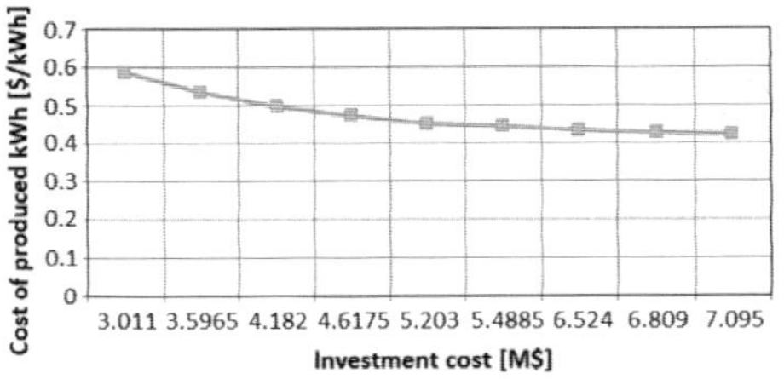

Figure 10: Variation of the cost of produced kWh with the investment cost

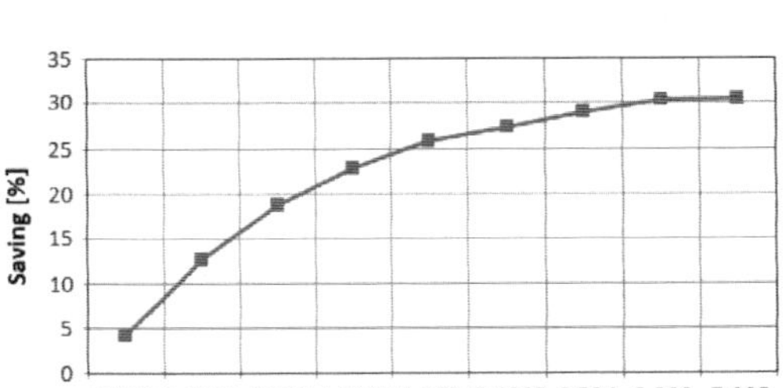

Figure 11: Variation of the saving with the investment cost

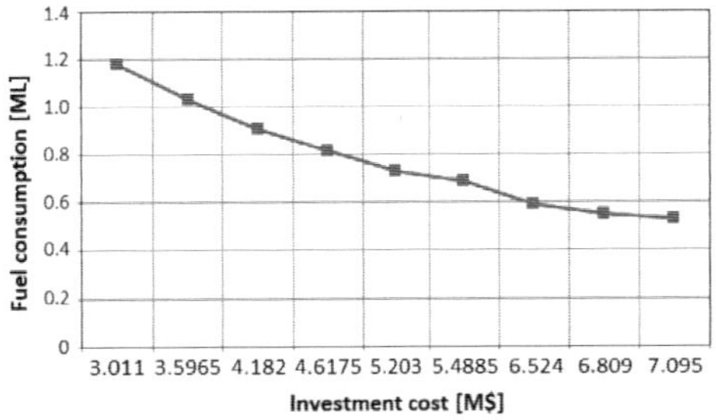

Figure 12: Variation of the fuel consumption with the investment cost

The curves shown in figures 10 and 11 are characterized by two sections. The first is transitory and the second is quasi-linear. During the transitory phase, the cost of kWh and saving vary widely with the investment cost (IC), but beyond this phase, the variation becomes minimal (at very high investment). Thus, it would be wise to choose the number of wind turbines in order to correspond to the transitional phase where a compromise can be obtained between a best saving and minimal costs of investment and produced kWh.

Other important information is the annual fuel consumption of the diesel generators. As shown in figure 12, the fuel consumption decreases quasi-linearly with the investment cost. Then, the ratio (fuel consumption/investment) is indifferent compared to the results presented in the figures 10 and 11. We need to increase the installed wind power capacity if we want to reduce the use of the diesel generators. A compromise must, also, be founded between saving of fuel consumption and cost investment.

VI. CONCLUSION

The renewable energy systems have intermittent output characteristics and are integrated with conventional power sources to deliver a stable power output. In various hybrid energy systems, diesel genset (DG) acts as steady source of power. The DG systems are designed to supply the load and also charge the battery energy storage system, if the renewable energy source along with battery is unable to supply the load.

A particular objective of this study was to show that where wind and solar resources are available, this system is able to developing and acting specifically in remote areas.

It was shown from the simulation results that the obtained optimal configuration of a hybrid energy system could overcome the effect of the meteorological and logistic conditions on the reliable supply of the village.

In the case of Tuktoyaktuk village, the simulations results prove that the best solution to reliable load supply without interruption and with minimum investment cost is the combination of a diesel generator with the hybrid wind-battery system (WDBHS) used ten wind turbines of Enercon-E33 and two diesel generator. This combination allows a fuel saving about 30%.

REFERENCES

[1] S. Rehman, I.M. El-Amin, F. Ahmad, S.M. Shaahid, A.M. Al-Shehri, J.M. Bakhashwain, A. Shash,2007. Feasibility study of hybrid retrofits to an isolated off-grid diesel power plant..

[2] L. H Tay, W. W. L. Keerthipala and L. J Borle, " Performance analysis of a wind/diesel/battery hybrid power system", Australasian Universities Power Engineering Conference, Perth, Australia, 23-26 Sept. 2001, pp. 107 - 112.
[3] N. Mendis, K. Muttaqi, S. Sayeef and S. Perera, "Power generation in isolated and regional communities: Application of a doubly-fed induction generator based wind turbine", 19th Australasian Universities Power
[4] N. Mendis, K. M. Muttaqi, S. Sayeef, S. Perera. Operation of a Wind-Diesel-Battery based Hybrid Remote Area Power Supply System. 6th International Conference on Electrical and Computer Engineering ICECE 2010, 18-20 December 2010, Dhaka, Bangladesh.
[5] Pragya Nema, R.K. Nema and Saroj Rangnekar, 2009. A current and future state of art development of hybrid energy system using wind and PV-solar: A review.
[6] Renewable and Sustainable Energy Reviews, 13(8):2096- 2103.
[7] Hongxing Yang, Zhou Wei, Lou Chengzhi, 2009. Op=mal design and techno-economic analysis of a hybrid solar– wind power generation system. Applied Energy, 86(2):163–169.
[8] Wei Zhou, Hongxing Yang, Zhaohong Fang, 2008. BaNery behavior prediction and battery working states analysis of a hybrid solar–wind power generation system. Renewable Energy, 33(6):1413–1423.
[9] S. Diaf, M. Belhamel, M. Haddadi, A. Louche, 2008.Technical and economic assessment of hybrid photovoltaic/wind system with battery storage in Corsica Island. Energy Policy, 36(2):743–754.
[10] Ali Naci Celik, 2003. Techno-economic analysis of autonomous PV–wind hybrid energy systems using different sizing methods. Energy Conversion and Management; 44(12):1951–1968.
[11] E. Muljadi and J.T. Bialasiewicz,"Hybrid power system with a controlled energy storage", IEEE Ind, Electron. Soc. Annu. Meeting Conf, Virginia, USA, vol. 2, 2-6 Nov. 2003, pp. 1296 - 1301.
[12] E. Muljadi and J.T. Bialasiewicz,"Hybrid power system with a controlled energy storage", IEEE Ind, Electron. Soc. Annu. Meeting Conf, Virginia, USA, vol. 2, 2-6 Nov. 2003, pp. 1296 - 1301.
[13] L. Wei and G. Joos, "A Power Electronic Interface for a Battery Supercapacitor Hybrid Energy Storage System for Wind Applications", Power Electronics Specialists Conference, Rhodes, Greece , 15-19 Jun.
[14] Murdoch University, 2009.
[15] HOMER.
[16] A. Salmani, S. Sadeghzadeh, M.R. Naseh. Optimization and Sensitivity Analysis of a Hybrid System in Kish-Iran. International Journal of Emerging Technology and Advanced Engineering. Volume 4, Issue 1, January 2014.
[17] N. Mahmud and a. Hassan "Modeling and Cost Analysis of Hybrid Energy System for St. Martin Island Using HOMER" Informatics, Electronics & Vision (ICIEV), 2013 International Conference
[18] Lambert, T., Gilman, P., Lilienthal, P. (2005) 'Micro-power System modelling with Homer', in Farret, F.A., Godoy Simões, M. (eds), Integration of alternative sources of energy, pp. 379 - 417
[19] R. Sen. Off-grid electricity generation with Renewable Energy Technologies in India: An application of HOMER. MSc. Renewable Energy & Environmental Modelling University of Dundee. September 2011.
[20] L. Olatomiwa, S. Mekhilef, A. S. N. Huda, K. Sanusi. Techno-economic analysis of hybrid PV–diesel–battery and PV–wind–diesel–battery power systems for mobile BTS: the way forward for rural development. Energy Science & Engineering published by the Society of Chemical Industry and John Wiley & Sons Ltd. March 2015.
[21] Trojan Battery Company.
[22] G. El-Jamal, M. Ghandour, H. Ibrahim, A. Assi. Technical feasibility study of solar-pumped hydro storage in Lebanon. IEEE, International Conference on Renewable Energies for Developing Countries 2014 (REDEC 2014), pp. 23-28, Beirut, Lebanon, 26-27 Nov. 2014.
[23] G. El-Jamal, H. Ibrahim, M. Ghandour. Investigation of the Technical-Economic Contribution of Renewable Energy and Energy Efficiency: Lebanese Context. IEEE, International Conference on Renewable Energies for Developing Countries 2014 (REDEC 2014), pp. 174-179, Beirut, Lebanon, 26-27 Nov. 2014.
[24] G. El-Jamal, M. Ghandour, H. Ibrahim. Application of multi-criteria decision-methods for the selection of renewable energy sources, Lebanese case study. 2nd International Conference Efficient Building Design: Materials and HVAC Equipment Technologies, Beirut, Lebanon, September 22-23, 2016.
[25] G. El-Jamal, H. Ibrahim, M. Ghandour. Integration of Energy Storage in a Wind-Diesel Hybrid System - Techno economical & Operational Advantages. International Conference on Renewable Energies for Developing countries (REDEC), Beirut, Lebanon, 13-15 July 2016.

Enhancing Photovoltaic solar panel

Raising efficiency of photovoltaic solar panel by preventive actions

Georges GEAGEA, Abdallah BATACHE, Henri EL ZAKHEM
Department of Chemical Engineering
University of Balamand,
P.O. Box : 33, Amioun, El Koura, Lebanon
Chemicalengineering@balamand.edu.lb

Marie-Thérèse MOUFAREJ ABOU JAOUDE
Department of Process Engineering
ISSAE - Cnam Liban
Avenue of the Cité Sportive
P.O. Box: 113 - 6175 Hamra
1103 2100 Beirut - Lebanon
mtaboujaoude@isae.edu.lb

***Abstract*—this document describes the main factors responsible for the reduction of the efficiency of photovoltaic (PV) solar panel. Those factors are: type of material used, accumulated dust on solar panel, higher temperature, position of the panel, and low area for photon capturing. To achieve higher efficiency, this paper investigated several ways to reduce the effects of the affecting parameters: reducing the temperature of the PV panel, eliminating the dust, controlling the position of the panel and adding a mirror to collect more photons. Those modifications were applied on a laboratory-scale prototype in order to enhance the performance of the (PV) to deliver higher efficiency.**

Keywords— efficiency of PV; Photovoltaic solar panel; renuable energy.

I. INTRODUCTION

Electrical energy forms a major source of comfort to human. Numerous activities would have been impossible without the aid of electrical energy. Without it, our current living standard cannot be maintained. In order to solve the issue of energy crisis facing the world nowadays, conversion of solar energy into electricity is introduced as a technique to be applied [1]. There are many types of the solar cells. Serious efforts are ongoing to improve their performance to the stages of practical applications [2].

Different techniques and processes have been applied through the years to improve the performance of the photovoltaic solar system and thus, increasing its efficiency. This paper aims to study the different parameters affecting the performance of photovoltaic (PV) solar system. Some modifications in the parameters were examined for this reason: type of material used, the environment which is a major contributing factor that includes the external resistances, and the geographical locations [3]. Recommended to achieve higher efficiency, more innovative design is presented in more advanced technique, taking into consideration solutions to the limitations of certain of the parameters that may affect the system's performance.

II. INFLUENT PARAMETERS AFFECTING THE PV EFFICIENCY

A. Material Vs. Efficiency

1) Selecting cells type

Selecting the chemical material is the main importance for higher productivity with less cost. Table 1 summarizes the top materials used and their characteristics.

TABLE 1: EFFICIENCIES OF SOLAR CELLS: ONLY CELLS WITH A CELL SURFACE LARGER THAN 1 CM² HAVE BEEN CONSIDERED [4].

Material	Type	Lab efficiency %	Manufac efficiency %
Silicon	monocrystalline	24.7	14.0-18.0
Polysilicon simple	polycristalline	19.8	13.0-15.5
MIS inversion layer (silicon)	monocrystalline	17.9	16.0
Concentrator solar cell (silicon)	monocrystalline	26.8	25.0
Silicon on glass substrate	Transfer technol	16.6	
Amorphous silicon, simple	Thin film	13.0	8.0
Tandem 2 layers, amorphous silicon	Thin film	13.0	8.8
Tandem 3 layers, amorphous silicon	Thin film	14.6	10.4
Callium indium phosphate	Tandem cell	30.3	21.0
Cadmium-telluride	Thin film	16.5	10.7
Copper indium di-selenium	Thin film	18.4	12.0

2) Choosing the right Specifications of the material.

Crystalline Silicon is usually used for photovoltaic cells due to the fact that Silicon is a semiconductor with the largest market penetration, it has been theoretically understood, and is the simplest to deal with [4]. On the other hand, amorphous - Silicon is cost-efficient and has good absorption features which allow it to be used as a base material for a thin solar cells film. Yet in the end Crystalline Silicon dominates as base material [5].

Regarding the high cost of the material by which the PV is made, polycrystalline type material is used in the design, which is not expensive and it provides close properties of the advanced PV materials, to perform many experiments.

B. External resistances

1) influence of dust

The dust accumulation on the PV's surface, the bird droppings or the water stains (salts), can play an important role of obstruction or distraction of the light energy to reach the solar cells, and this minimizes the efficiency [3]. All these types of external resistances could be referred to as "soiling". The surface of the PV is a major contributing factor in the soiling phenomenon. If a PV surface is rough, sticky, furry and not smooth, that would provide the best conditions for dust accumulation.

Describing the phenomenon of dust deposition is a must. By definition, the dust is a term applied to solid particle having a diameter less than 500 µm. It is present in the air and occurs in the atmosphere coming from many sources like dust lifted up by wind, vehicular movement, volcanic eruptions, and pollution [3]. There are two primary factors concerning the dust settlement on PV systems: the qualities of the dust and the indigenous surroundings. The property of dust that includes the type, the chemical - biological and electrostatic property, in addition to the size, shape and weight, is at equal importance of its accumulation or aggregation. It is also very known that the dust accumulations promote dust. In other word, with the initial onset of dust, it would have the tendency to attract or promote further settlement. Which means that the amenability of the surface to dust collection increases [6].

As for the local environment, it implicates site-specific factors affected by the nature of human activities, built environment characteristics such as surface finishes, orientation and height of installation, vegetation type and weather conditions (the geographic location is at major importance in the section) [7].

Thus, the purpose of this work is to access the impact of the accumulation of dirt on the efficiency of solar PV panels. The study was operated indoor. It involved artificial lighting, instead of the sunlight, by mean of spotlights to neglect the variation that may be occurred under the sunlight.

As a solution: a proper maintenance for the solar panels would be of a high importance. Some particles like dust and sand can be eliminated when washed away. Washing can be done naturally in the winter season when it rains, but this is not efficient for a long term since the seasons change through the year. For this reason, there is a need for water to provide a permanent cleaning of the PV surface and thus ensuring an optimum power generation.

To solve this problem, two techniques could be applied: the manual cleaning or mobile cleaners.

The manual cleaning method follows the same concept as cleaning the windows of a certain building. To prevent the scratches of the module, scrubbing of the soil off from the surface and brushing it with special bristles were both applied. It must be mentioned that this method requires a long ladder and a scrub [7].

When it comes to the mobile cleaners, this technique uses machinery to perform the mission in addition to a water storage supply to clean the PV's surface [7].

Thus, to mitigate the impact of dust, a cleaning cycle must be recommended. This is why, in this design, a spraying water system made of pressurized water nozzles was used to remove the accumulation of dirt and dust especially in the dry seasons and in the case of intensive dust accumulation. It is a simple and a cheap solution.

2) Influence of shading

Some of the patches caused by the deposition of leaves and other dirt that covered some of the PV cells reduce the energy production. They have a sever effect on PV performance. The *figure 1* below shows a PV model composed of 10 successive cells in series with one shaded cell among them [3]. The shaded cell is considered as a resistance that blocks the flow of the current generated from the other cells, so it is unable to produce electricity like the others. The problem is that dust stops the current cell from producing energy. Also it blocks energy produced by previous cells from passing through this cell. This fact causes a heat up of the shaded cell and leads to the occurrence of a hot spot that will ultimately destroy the system.

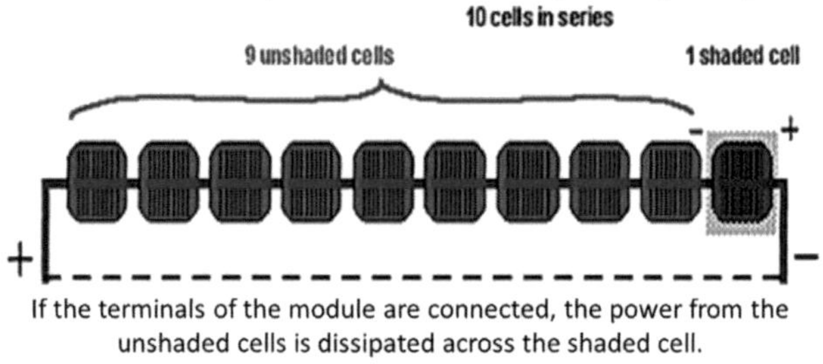

Figure 1: Current Flow across the Shaded Cell [3].

In order to minimize the impact of partial shading on PV, bypass diodes could be used especially when a single unit is involved in a series string. It must be mentioned that when dealing with a sequence of cells connecting in series, the same amount of current flow through every single cell.

Thus, in a similar case of the one represented in *figure 1*, the shaded cell will not be able to produce an amount of current equivalent to the unshaded cells' current.

Two options are possible for this problem:

The solution would be by inserting a bypass diode between each 4 cells, this diode will by-pass the shaded cell, the current will pass around the shaded area, reducing by that the voltage losses through the module. *Figure 2* shows that if the partial shading takes place, the current produced from the unshaded cells passes across the bypass diode rather than the shaded cell. The most important advantage gained is that the amount of local heating at the shaded area will be drastically reduced by the use of bypass diode, preventing the formation of hotspots. This is a

https://doi.org/10.1051/matecconf/201817101005

good solution but presents several inconveniences: it is expensive, and it doesn't prevent shading, it only bypasses shaded cells. This technology could not been used for this work, but it will be presented in another paper.

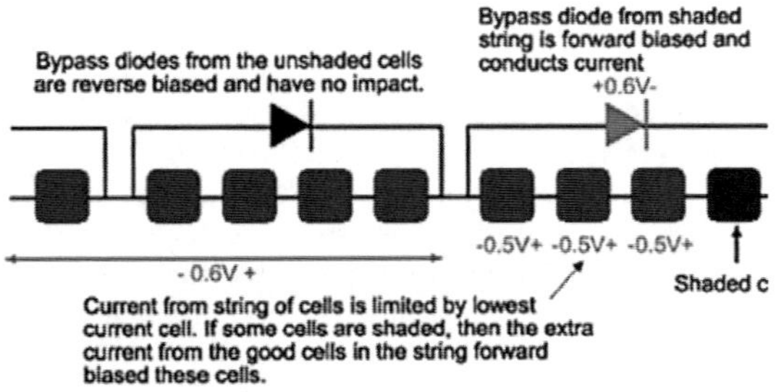

Figure 2: Introduction of Bypass Diode in PV Modules for Hot Spot Avoidance [3]

3) Effect of the temperature

Temperature is one of the most important factors that must be taken into consideration when dealing with a PV design. Examining the effect of the ambient temperature on the efficiency of a solar photovoltaic panel is a must. Recently, there has been a huge increase in the understanding of the operational system of photovoltaic apparatus, leading to an increase in the power conversion efficiencies of such devices. Solar cells vary with the variation of the temperature because it will affect the power and the voltage output from the cells [8]. Therefore, PV panel is influenced by varying weather conditions. Since the weather is not at constant permanent conditions, and the solar panels are installed all over the world in different climate regions, most solar panels are not operating under ideal conditions. This is why, it is crucial for engineers to know the operating system of the panels and how they react to different weather conditions [9].

Temperature affects the electron activity, which means how electricity can flow through an electrical circuit by changing the speed needed by the electrons to travel. It must be cited that an increase in the temperature is a result of an increase in resistance of the circuit. Thus, resistance is decreased with decreasing temperatures which leads to higher efficiency [10].

As it was mentioned previously, the power output from the cells is affected by the change of the temperature. Which means that the voltage depends on the temperature change: an increase in the temperature leads to a decrease in the voltage. The results obtained by Vahid *et al.* (*Figure 3)* show the variation of the current vs. the voltage at different temperature values [9]. It could be noticed that the output voltage of the PV increases when operating the cell at a lower temperature. A similar interpretation applies for the power delivered *(figure 3),* the output power of PV module increases with decreasing temperature.

Needless to say that the power output from the cells is affected by the change of the temperature: an increase in the temperature leads to a decrease in the voltage [9]. It must be cited that the efficiency differs by changing the material of the PV since the efficiency of various materials has varied levels of dependence on temperature. This is why, a PV system must be also engineered based on an understanding of the materials used in the PV panel [4].

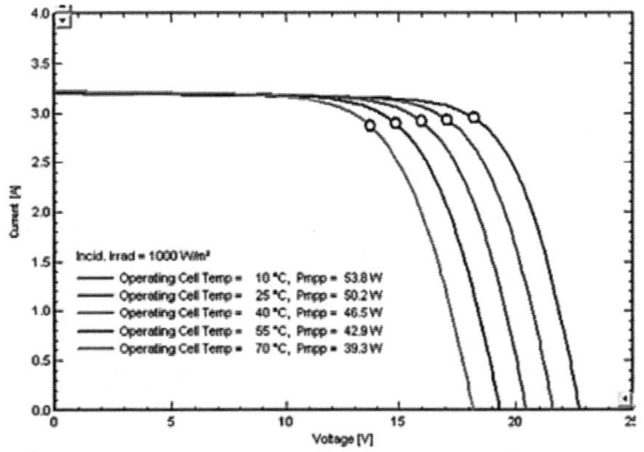

Figure 3: Output current and voltage of the PV module at different temperatures [3].

In this design, the work is done by keeping the panels as cool as possible by introducing an active cooling system by pumping water behind the panels to pull the heat away. The increase in the temperature of the cooling fluid depends on the flow of water sent through the pump.

4) Effect of the PV position

One of the main causes for low efficiency in PV cell is that the PV solar panel is putted in a fixed angle on the roof, and that the light coming from the sun is not sufficiently intense. The angle of the solar panel has a very important role in increasing the output current. That's why the efficiency changes between the day hours and differ between different months.

Putting many solar panels may be a solution to collect more photons and it will function. But it is not feasible because of the high cost of PV solar panels. So an innovative solution is needed to makes the solar panel collect more photons in less cost possible.

In this design, to solve the problem of the low collection of photons, a mirror was added facing the panel. The introduction of the mirror, which is considered a cheap material, increases the collection of photons and provides more energy. Also, as a solution for the sun position changing, a screw is put on the back of the panel and on the back of the mirror. In this way, the position of the panels will be modified to capture the maximum amount of energy coming from the sun.

III. Experimental design

A. How this design helps solving efficiency problem

In the design, several preventive solutions were introduced for the main "efficiency decrease" in a normal PV solar panel. Figure 4 shows a realistic photo of the invented prototype. It has the shape of an open book in which in one of its sides the photovoltaic cells is fixed and, on the other side, the mirror is fixed. When it is closed, it looks like a box having a certain width. Both of the sides are surrounded by wood frames. The wood is coated by paints to prevent its corruption in the case of water leakage. The mirror is used to increase the sunlight absorbance.

The photovoltaic solar system is made up of silicon since it was proved that silicon has provided the maximum achievable efficiency up to these days.

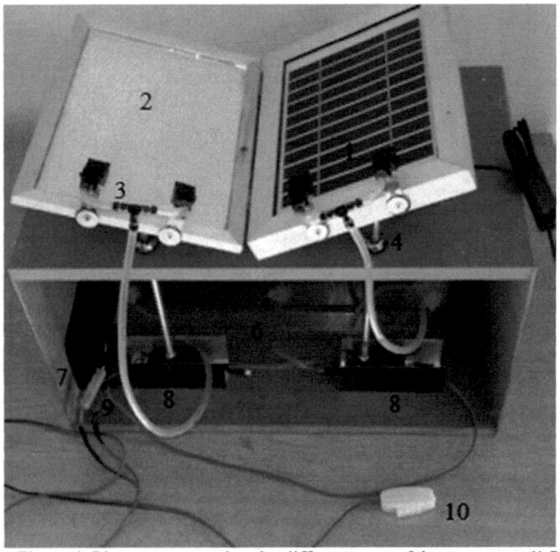

Figure 4: Picture representing the different parts of the prototype: 1) PV solar panel 30cm*15cm, 2) Mirror; 3) Nozzle; 4) Screw to control the PV angle 5) Connection tubes; 6) tank; 7) battery; 8) pumps; 9) protective fuse; 10) switch.

Removing the dust accumulated on the PV surface is of a major importance, that's why a spraying water system is provided to get rid of dust settlement. The spraying system is made up of a sequence of nozzles fixed on the wood frame on both sides: the PV side and the mirror side. Its responsibility is to clean the system surface in the case of dust accumulation. Those nozzles have been fixed on movable screws where they can change the direction of the water split. The nozzles are connected by pipes made of PVC to two powered pumps present in the box under the PV surface side. They will remove the dust by the pressure water jet system. By applying this technique, another important factor is included: cooling. The water has the ability to cool a system four times more than the air. Thus, the water here has double functions: cleaning and cooling.

Controlling the angle by which the PV system should be set is another contributing factor. It is meant by the angle the degree of inclination with reference to the ground surface. For this reason, two screws are implemented, they are attached to a railway to fix the position of the solar panel and the mirror, according to the direction of the sun, in a way that they can collect the maximum quantity of photons.

A cooler in the back of the panel is introduced to reduce the effect of the heat by the sun. The cooler was made of galvanized metal because it has a good heat transfer coefficient and is easy to be shaped on a small scale. The cooler has an "S" shape water canalized pipes for a better cooling configuration design and the tank has a volume of 275ml.

Also the reflecting light is being prevented by introducing an anti-reflective layer at the bottom of the PV solar panel. This will help the 30 % of light that will be reflected by default to be captivated by the solar cells and used instead of being dispersed. Noting that the water won't be considered as a reflecting agent, because the width of the water layer is small enough and won't be always present on the top of the protective coating layer.

Finally the battery used in the system is a GT4B-5, AH: 2.3, size: 4.4 x 1.5 x 3.2 cm and it is made in China.

It must be mentioned that the experiment was done in a laboratory where the light source used was a 30 Watts Neon lamp. Several measurements of the output current (mA) and voltage (V) vs. Time (minutes) were applied on two systems to compare their performance: a normal panel put one week outside on the balcony in El Koura, North of Lebanon (PV_1) and our prototype (PV_2).

IV. EXPERIMENTAL RESULTS

In this study, preventive actions were done for the main factors that enhance the resistance of the solar cell: by decreasing the shading effect by dust, and putting an anti-reflective layer on the top of the solar panel, and putting nozzles to eliminate dust, and increasing the photons quantity absorption.

The voltage was measured and given by the prototype (PV_2) and compared it to the normal panel (PV_1). The figure 5 shows that the prototype designed is more efficient since it delivers higher voltage range and the normal PV panel has a lower voltage.

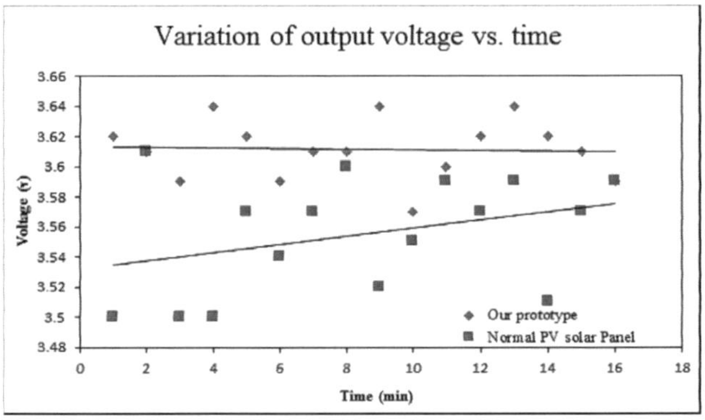

Figure 5: Variation of the Output Voltage of both PV_1 (prototype) and PV_2 (normal PV solar Panel) vs. Time

Another experiment is introduced in order to compare the current charging power of a phone's battery from different electrical sources. Four sources were used to charge a Samsung Galaxy grand 2 cell phone as it is shown in the *figure 6* below: charging from a socket charger, from the normal PV, from the prototype, and when it is unplugged. The purpose of this experiment is to compare between the electricity produced from the two system PV_1 and PV_2 when it is needed to charge a cell phone.

This measurement was done using an application called battery graph from the play store. It measures how the battery is being used and how much energy is being lost or added during time (figure 6).

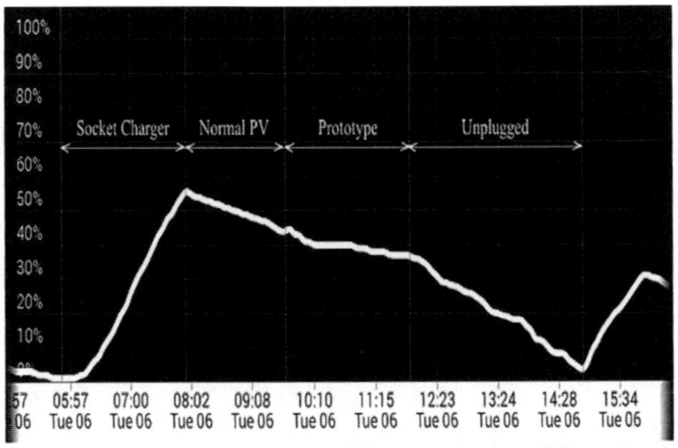

Figure 6: The current variation curve of the phone battery from different electric resources

The first section of it represents the current variation curve of a phone charging from the wall socket, the second section shows the current variation curve of a phone being charged by the PV_2 using a neon lamp of 30 watts, the third section indicates the current variation curve of a phone charging by the electricity produced from the prototype designed (PV_1) with the same neon lamp, and the last section represents the variation curve of a phone when it is unplugged.

In *figure 6,* when the curve decreases, it means that the battery is consuming electricity, and thus the phone is consuming more than it is charging. When the curve increases, it means that the phone is charging. When the phone is plugged in a socket charger, the curve starts to increase which means that the phone is charging. It takes electricity more than consuming to supply its usage. When the phone is charged by the electricity produced from the PV_2, the curve decreases slowly which means that even the system is charging the phone but its consumption is higher than the electricity delivered to it from PV_2. Moreover, when the phone is charged from the electricity produced from the prototype PV_1, an approximately constant line appears in the graph which means that the electricity received from the prototype is equal to the electricity needed for the phone to function normally. Finally, when the phone is unplugged from any source of electricity, the curve decreases rapidly indicating that the phone is not charged anymore and it consumes all its stored electricity.
Needless to say that the slope in the section of the normal PV_2 is decreasing, but the slope of the curve resulting from the prototype is almost constant, with small variation. This experiment shows that the electricity produced from the prototype is more efficient than that from PV_2, so the purpose of the experiment is verified.

From those two graphs, we can conclude that if those changes were done on a larger scale, this will increase the production of energy and decrease the resistivity in a feasible cost.

V. CONCLUSION

The results obtained from the experiments examined on the advanced prototype show that all the changes added to the new design enable it to increase the production of energy and to decrease the resistivity effects.

In the future works, the water flow rate must be optimized to have the best cooling of the photovoltaic panels. The latter should be studied while optimizing the energy consumption of the cooling system.

Furthermore, the quantum dot technology can be implemented in the prototype. This new technology would not only help to eliminate the resistances, but also to increase the capacity of the cell to take more wavelength and benefit of the entire solar wavelength.

REFERENCES

[1] Barnes, B., & Fulford, G. R. (2014). *Mathematical Modelling with Case Studies: Using Maple and MATLAB* (Vol. 25). CRC Press.
[2]: Jelle, B. P., & Breivik, C. (2012). *State-of-the-art building integrated photovoltaics. Energy Procedia*, *20*, 68-77.
[3]: Mohammad, R. M., Hashim, H., Chandima, G., Mohd, A. R., Mohammad, I. R., & Shahrooz, H. (June 2016). *Power loss due to soiling on solar panel: A review. (59), 1307-1316.*
[4]: Kaltschmitt, M., Streicher, W., & Wiese, A. (Eds.). (2007). *Renewable energy: technology, economics and environment*. Springer Science & Business Media.
[5]: Li, G., Shrotriya, V., Huang, J., Yao, Y., Moriarty, T., Emery, K., & Yang, Y. (2005*). High-efficiency solution processable polymer photovoltaic cells by self-organization of polymer blends. Nature materials*, *4*(11), 864-868.
[6]: Goossens, D., & Van Kerschaever, E. (1999). *Aeolian dust deposition on photovoltaic solar cells: the effects of wind velocity and airborne dust concentration on cell performance. Solar Energy*, *66*(4), 277-289.
[7]: Shaharin, A. S., Atul, K. S., Mior, M. M. M., & Mohammed, A. B. R. (2014). *Influence of Dirt Accumulation on Performance of PV Panels*, (50), 50-56.
[8]: *Photovoltaic Efficiency: The Temperature Effect.* Lesson 2 - Fundamentals Article (n.d).
[9]: Vahid, J. F., Majid, D., & Jafari, F. (November 2011). *The Effect of Temperature on Photovoltaic Cell Efficiency.*
[10]: T.T. Chow, (February 2010). *A review on photovoltaic/thermal hybrid solar technology. (87), Issue 2, 365-379.*

Optimization of Energy Management of a Microgrid Based on Solar-Diesel-Battery Hybrid System

Hussein Ibrahim
Institut Technologique de Maintenance Industrielle (ITMI), Cégep de Sept-Îles, 175 rue de la Vérendrye Sept-Îles (Québec) G4R 5B7, Canada
hussein.ibrahim@itmi.ca

Mazen Ghandour
Faculty of engineering III, Lebanese University, Hadath-Beirut, Lebanon
drmazengh@yahoo.com

***Abstract*— Hybrid systems, which are composed of combinations of diesel generators, battery energy storage system and renewable energy resources such as photovoltaic, are outlined as a recommended approach for off grid power supply options for remote areas applications. Since these systems are not connected to an infinite source of energy, they must be well designed and controlled to satisfy the demand load. This study presents an efficient battery management strategy for the charging and discharging of the batteries in a hybrid renewable energy systems by controlling the energy flow between different components of the system. In order to simulate the developed battery management strategy, the components of the hybrid system are studied and modeled using mathematical models. During modeling of the battery storage system, the effects of aging and capacity degradation were taken into consideration. The simulation of the strategy was based on a case study, where it is validated to be functional. Finally, the optimization of the strategy has been done by working on its critical parameters.**

***Keywords*— Hybrid renewable energy systems; energy storage; battery; PV; energy management strategy; optimization.**

I. INTRODUCTION

Public interconnected power grids are composed of complex combinations of generation plants, substations, transformers and transmission lines, which supply electricity to cities, businesses and industry. In addition, there are smaller independent power grids that provide power to islands or remote areas, which have limited or no access to public interconnected grids. Traditionally, small stand-alone grids are electrified by diesel generators. However, the renewable energy resources are attractive sources of power, since they can provide sustainable and clean power. Hybrid plants can be an integration of diesel generators with renewable energy resources such as photovoltaic. Because of its intermittent and irregular nature, PV generation makes hybrid system management harder. Consequently, for some authors [1-2], PV production into the grid is considered to be limited [3]. One of the major challenges for PV systems remains in the matching of the intermittent energy production with the dynamic power demand. A solution is to add a storage element to these nonconventional and intermittent power sources [4-5]. In this case, the hybrid system is composed of a PV generator, local loads, electricity storage, and the conventional source (diesel genset) [5]. In addition, integrating a battery energy storage system with the hybrid plant provides significant dynamic operation benefits such as higher stability and reliability of power supply.

More recently, hybrid renewable energy systems (HRESs) are gaining popularity for their environmental friendliness and potential economic benefit. However, HRESs have not yet dominated in the market due mainly to their high price. One of the major factors leading for this high price is the high battery cost. Thus, manufacturers have been focusing on the reduction of battery cost to increase their market share. An effective battery management strategy (BMS) can reduce the required number of battery cells and extend the replacement period, hence reducing the battery cost. As part of our objective of designing such a BMS, we will in this study focus on extending batteries' life with respect to the whole investment life which has significant impact on the battery cost.

Battery life is defined as the duration (measured in the number of cycles or elapsed time) of a rechargeable battery until it degrades irreversibly and cannot hold a useful capacity of the underlying applications [5]. That is, the battery capacity monotonically decreases with time, and will never be recovered. Among the various reasons for battery capacity degradation, the discharge/charge rate is reported to be the most critical. For example, a continuous exposure to high discharge current leads to fast capacity degradation thus shortens the battery's life, which is unavoidable due to sudden changes in the power requirement, such as motor start-up. To extend the battery life, researchers focused on management strategies that use more than one charging/discharging modes.

In this study we will elaborate an advanced battery management strategy that best saves battery life-time respecting some constrains and system conditions. We will take a case study and size the hybrid system, so we can take it as a reference to simulate our strategy.

II. TOPOLOGY OF THE HYBRID SYSTEM

Before focusing on each component in the system, the overall topology must be defined in order to specify these components and define their role. A scheme of the studied system is presented in Figure 1. The different power directions are represented. The sign convention in Figure 1 is used as a reference throughout the whole paper. The main components of the hybrid system are the PV generator, the batteries' energy storage, the user loads and dump load, the diesel generator, and the electronics power converters. The parameters and, respectively, the values of the state of charge and state of health of the batteries, with calculations are detailed in following sections. According to the sign convention, the laws of physics require the power balance in the system described by:

$$P_{DIESEL}(t) = P_{LOAD}(t) + P_{DUMP\ LOAD}(t) - P_{PV}(t) - P_{BAT}(t) \quad (1)$$

With P_{DIESEL}, P_{PV} and P_{BAT} are the power supplied by the diesel and photovoltaic solar power plants and battery energy storage, respectively. P_{LOAD} and $P_{DUMP\ LOAD}$ are the power absorbed by the main load and the dump load respectively.

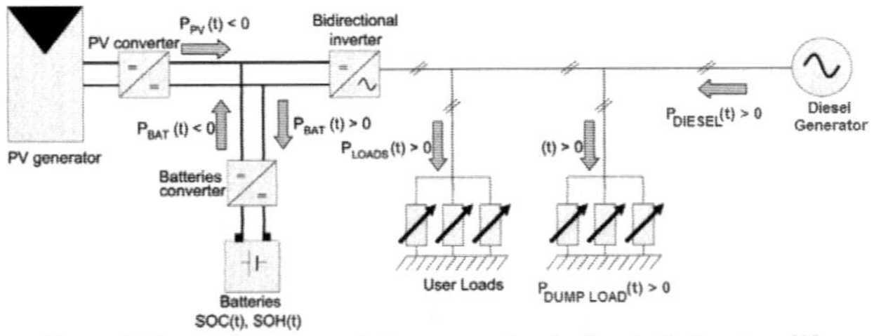

Figure 1: Power direction and sign convention in the studied system [3]

The main concept of this system is to generate electric energy from all possible sources and in any form (AC or DC and in any frequency) and transform them into appropriate voltage by using converters in order to supply storage system or loads. Many other sources could be added (Wind, hydro...) in the same concept to increase the quantity of collected renewable energy, but this would increase the system complexity and makes its study harder.

The battery bank will insure the autonomy of the system. It is equipped with a special DC/DC converter that controls the charging and discharging of the battery. The bidirectional inverter may control other components in the system in order to apply the best battery management strategy.

III. Energy conversion modeling of hybrid system

A. PV system

The method selected to model the PV array, in this study, involves simply using mathematical equations to calculate the current and voltage generated at the specified temperature and irradiance from the power generated at Standard Temperature Conditions (STC) by analyzing the factors concerned and determining how they would affect the power generated. STC is defined as $1000W/m^2$ irradiance, temperature of 25°C and air mass of 1.5 (corresponds to 48.2° incidence angle). If the charge controller includes a MPPT system, the PV panels always work at the maximum power point of the I = f (V) curve, so the output current of the PV generator is as follows [6]:

$$I_{pv}(t) = I_{mp} \cdot \frac{G(t)}{G_{STC}} \cdot \left[1 + \frac{\mu_{I,OC}}{100}(T_c(t) - 25)\right] \quad (2)$$

Where I_{mp} is the operating current at maximum power point and is given in the datasheet of the PV panel. Also the output voltage of the panel with MPPT controller is:

$$V_{pv}(t) = V_{mp} \cdot \left[1 + \frac{\mu_{V,OC}}{100}(T_c(t) - 25)\right] \quad (3)$$

Where $\mu_{V,OV}$ is the temperature coefficient of V_{OC} [%/°C]. Therefore, the resulting output power of the PV system with MPPT is:

$$P_{pv}(t) = V_{pv}(t) . I_{pv}(t) \quad (4)$$

B. Diesel generator

In this study we are studying the efficient power flow control and management to realize the best strategy that reduces the operational costs of the hybrid system. Since the diesel generator is an automated complete system that gives the demanded output power with stable voltage and frequency values, the internal function and composition of the system may not be our major interest. The most important issue that we must consider is the fuel consumption of the power generator and its relation with the demand load.

Figure 2 shows the diesel fuel consumption of a 15 KVA – single phase generating set as per manufacturer datasheet [7]. The relation between the hourly fuel consumption and the percentage of nominal load on the generator is nearly linear. Using linear interpolation methods, this graph could be used to model the generator and calculate its fuel consumption over the hours of operation.

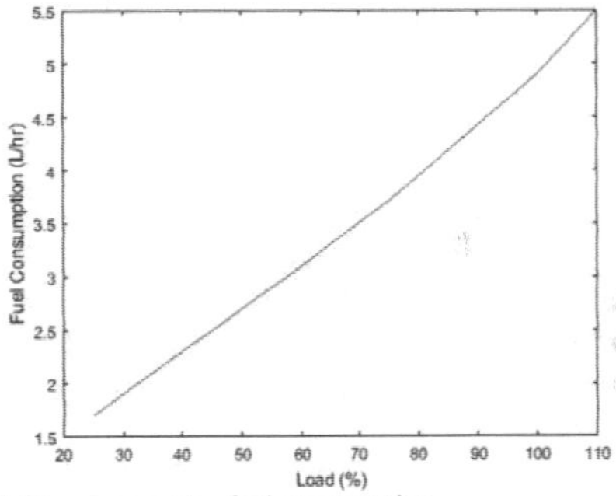

Figure 2. Diesel generator fuel consumption

C. Battery

a) Modeling of the battery

The model chosen to be used in this study is a model that takes into account the state of charge of the battery. It is a dynamic model that adopts the simple model where the battery is represented by an open circuit ideal voltage source and a fixed resistance (Figure 3), but with variable voltage source that depends on the SOC and constant internal resistance.

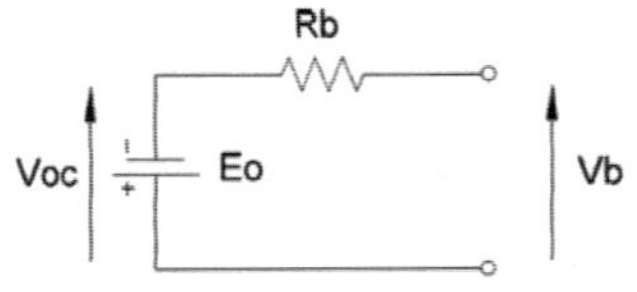

Figure 3: Simple Circuit Model [8]

The relation between the state of charge (SOC) and the open circuit voltage (Voc) is supposed to be linear as in the following equation [8]:

$$V_{OC} = U_{max}.SOC + U_{min}.(1 - SOC) \quad (5)$$

Where U_{max} is open circuit voltage at full battery and U_{min} is the open circuit voltage at empty battery, these values are taken from the battery manufacturer datasheet.

The charging current of the battery can be controlled by the charging voltage on the battery terminals. This current can be calculated as in the following equation [9]:

$$I_{ch} = \frac{V_S - V_{OC}}{R} e^{-a.SOC} \quad (6)$$

Where a is the coefficient that allows the adjustment of the equation, it can be calculated by substituting experimental values of current and state of charge.

The state of charge must also be modeled so our battery model is complete. The state of charge is changed in two cases, charging and discharging [10].

- During charging:

$$SOC(t + \Delta t) = SOC(t) + \frac{\eta_{bat}.I_{ch}.\Delta t}{C_b} \quad (7)$$

Where η_{bat} is the round-trip efficiency of the battery, it reflects the charging efficiency of the battery, since not all energy injected in the battery can be stored, certainly there will be losses. This factor has values between 0.7 and 0.85 during charging and is equal to 1 during discharging process. Δt is the charging time and C_b is the full battery capacity [10].

- During discharging:

$$SOC(t + \Delta t) = SOC(t) - \frac{I_{disch}.\Delta t}{C_b} \quad (8)$$

b) Cycling and Capacity Degradation

Modeling lifetime of batteries is a very important aspect of hybrid power system simulation because the uncertainty associated with the expected lifetime of the batteries makes the estimates of cost of energy of the projects very uncertain. Since the life cycle cost of the batteries is one of the significant power system expenses it is a major source of uncertainty for potential power system investors.

Many factors affect the life of the batteries, including the depth of the charge – discharge cycles, the current, the cell voltage, the performance of the charge controller (e.g., voltage and state of charge limits and regulation), the length of time that the batteries are in a low state of charge, the time since the last full charge, the temperature, etc.

Many studies have been published about the simulation and optimization of renewable stand-alone systems including batteries. However, the battery lifetime has always been estimated in fixed values based on the experience of many researchers or it has been estimated by calculating the number of equivalent full cycles. In the best cases, it may be estimated using the cycle counting method (CCM).

The CCM assumes that a complete cycle is reached the Ah-throughput is equal to the nominal battery capacity. The estimation of the lifetime consists of adding the charge (Ah throughput) cycled by the battery and calculating the number of full cycles (Z_N) as follows [8]:

$$Z_N(t + \Delta t) = Z_N(t) + \frac{I_{disch}.\Delta t}{C_N} \quad (9)$$

Where I_{disch} is the absolute value of the discharge current. C_N is the nominal battery full capacity. The end of life of the battery is reached when the degradation of the battery capacity reaches 80% of its nominal capacity. This capacity loss by degradation can be calculated from the Schiffer equation [8]:

$$C_{deg}(t) = 0.8\, C_N \,.\exp\left[-5.\left(1 - \frac{Z_N(t)}{1.6.Z_{IEC}}\right)\right] \quad (10)$$

Where Z_{IEC} is the total number of cycles estimated by the manufacturer according the IEC standards. Therefore, the remaining battery capacity is difference between the initial nominal capacity and the capacity loss by degradation:

$$C_{rem}(t) = C_N - C_{deg}(t) \quad (11)$$

IV. BATTERY MANAGEMENT SYSTEM

In order to elaborate, simulate and validate the management strategy, a data base is necessary. For this reason, we recovered the data of solar irradiance, ambient temperature, and of electrical load of the diesel engine on the experimental site situated at Canada (Latitude: +48.83, Longitude: -64.48) [10-11]. These data will be used to sizing the hybrid renewable energy system in order put a strategy compatible with it.

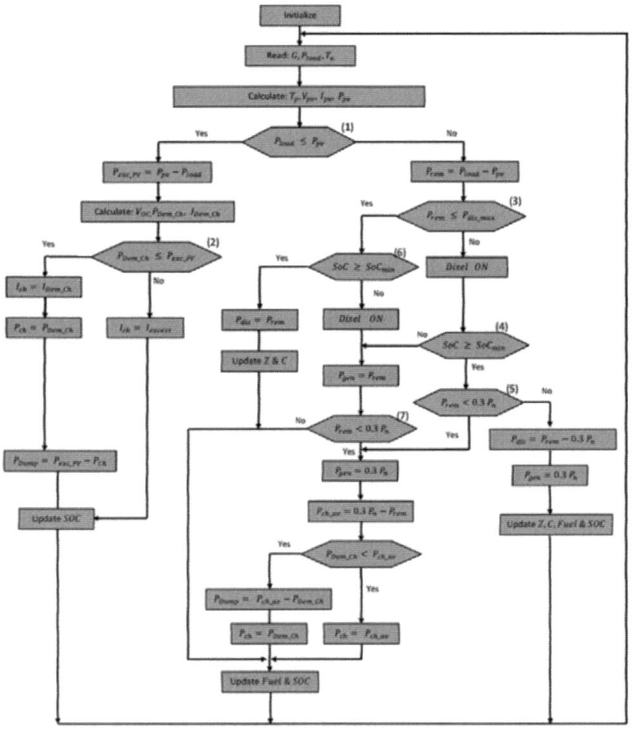

Figure 4: Energy management strategy flow chart

Taking into consideration the previously discussed constraints and priorities, and starting from the studied models of the system components, and taking the data of the case study as a reference, we elaborated the control and

https://doi.org/10.1051/matecconf/201817101006

management strategy of the PV-Diesel-Battery hybrid system. The flow chart in Figure 4 shows the strategy as an infinite loop that reads inputs from an external source each period of time and gives the outputs during this period. After the system parameters are initialized, the inputs must be read, which are the solar irradiation G, the demanded power by load P_{load} and the ambient temperature T_a. This data is taken from the case study profile table that gives a sample every 10 minutes. According to the mathematical model of the PV system, the temperature of the panel and the PV system output current and voltage are calculated. Losses in the converters are also taken into consideration. Based on these information, the power flows are correctly dispatched between the different sources (PV, diesel), loads (principal and dump loads) and battery energy storage in order to ensure the minimum cost per kilowatt-hour produced from the hybrid system and a long cycle-life of the batteries.

V. Results and Discussion

A. Validation of the energy management strategy

The input data of the load power consumption, the solar irradiation and the ambient temperature used to simulate and validate the management energy strategy, are given in a form of 10 minutes samples. The time samples are reduced to 10 minutes instead of 1 hour to have more accurate results. The strategy was validated on a two days input profile. Figure 5 to Figure 11 illustrate the obtained results.

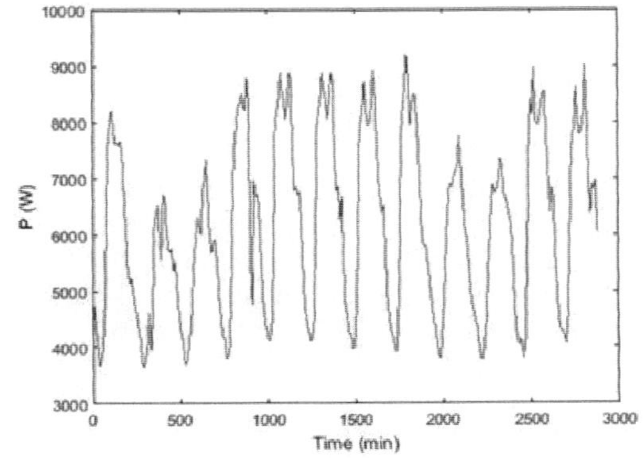

Figure 5: Variation of load power consumption

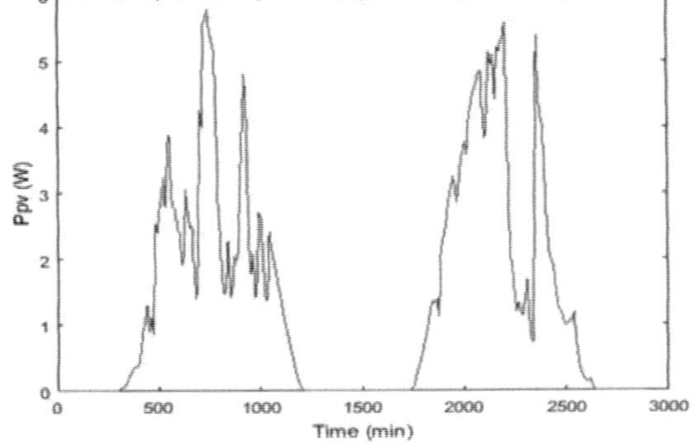

Figure 6: Variation of PV power produced

As we see in the Figures 5 and 6, load varies too much with time. The strategy must handle this variation well and satisfy the load. The following is the PV power produced; it seems to be very similar to the solar irradiation. This convergence is due to the absence of the effect of temperature, since this effect is very slight at temperature near 20°C.

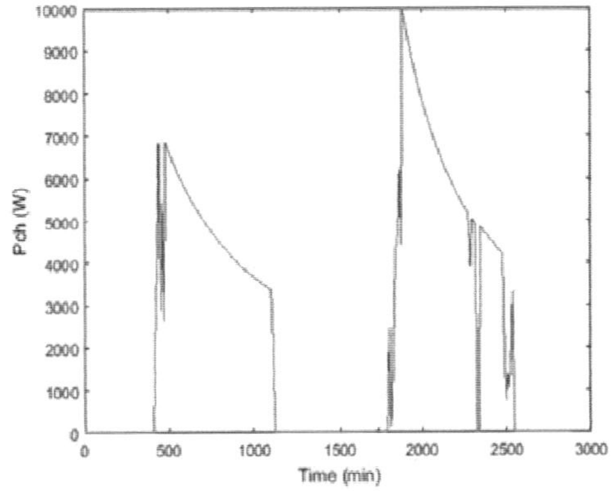

Figure 7: Variation of power charged into the batteries

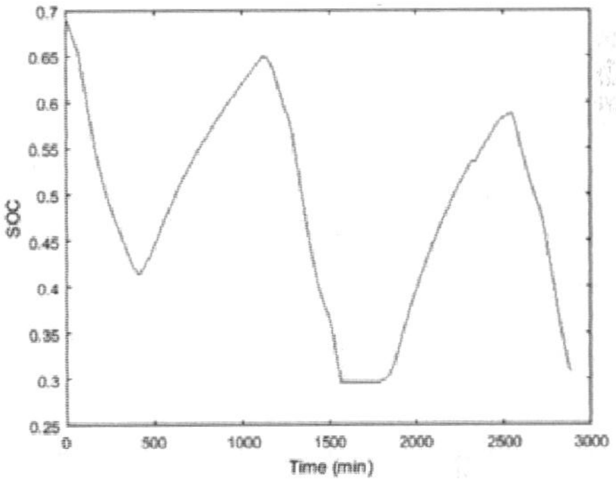

Figure 8: Variation of state of charge (SOC)

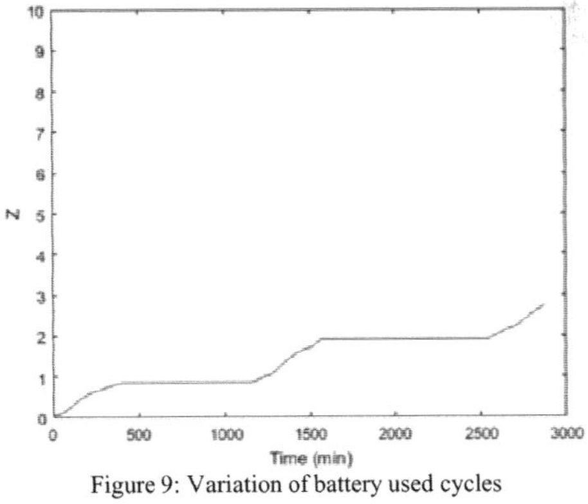

Figure 9: Variation of battery used cycles

Figure 7 shows the variation of the power charged into the batteries over the two days of simulation. As it is clear, the batteries are charging at times high solar irradiation, and the charging current (since the charging is done at constant voltage) decreases after the charging process starts, which is a true character of batteries and agrees with the charging characteristic given by the manufacturer [12].

Figure 8 shows the variation of the state of charge of the batteries. During the charging times, the SOC increases and during the discharging times it decreases. Also it is clear that there is a limitation of the SOC decrease to 0.3 or 30% of battery capacity.

Figure 9 shows the evolution of the consumed battery cycles. As the battery model previously described, the cycle counting is done during the discharge of the batteries, and this is clear in the graph. As we see also in Figure 18, the capacity of the battery is slightly decreasing, and this is a result of degradation due to cycling. Figure 10 and Figure 11 show the generator output power and the fuel consumed. They seem to be proportional, and this result was expected from the characteristic given by the manufacturer [7]. As it is clear the generator only turns on at the time where the state of charge reached 0.3 (between 1500 and 2000 minute), which is when the batteries are almost empty and there is no solar irradiation.

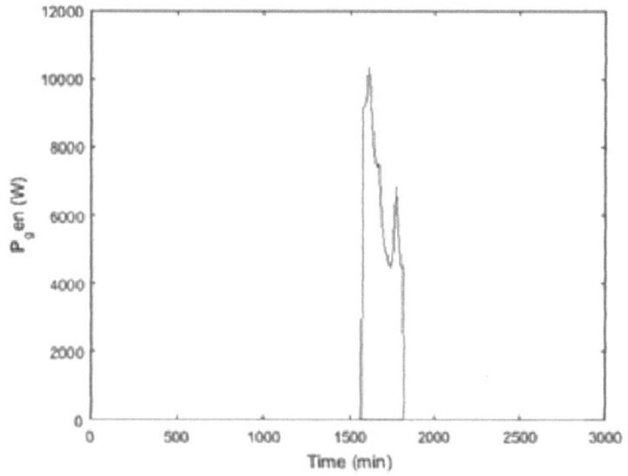

Figure 10: Variation of diesel generator output power

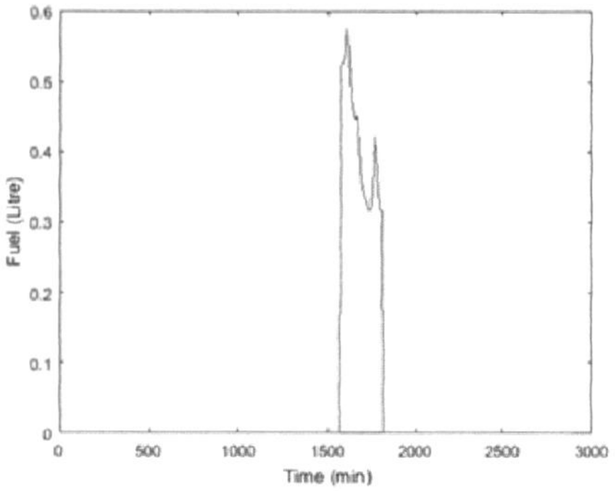

Figure 11: Variation of fuel consumed by diesel generator

All the preceding results are very logical and reasonable and this gives a validation of the work. It confirms that the strategy is working well and can be used to control any similar hybrid renewable energy system.

B. Optimization

During the optimization process we will add a new parameter that is to be calculated to evaluate the management strategy well and study it financially, which is the cost per hour of operation and maintenance of the system. The operating costs include the price of fuel consumed, the costs of generator maintenance and cost battery storage system over the total life time of batteries [13-19]. In order to fair compare strategies we will calculate the hourly operating cost of the system as an average of total life time.

The cost of diesel fuel in Canada is 1.3 USD per liter [13]. The generator oil must be changed every 250 operating hours [13]. The oil capacity of the generator is 10.6 liters; the average price of the liter is 7 US dollars [13], so the oil cost will be 10.6*7/250 = 0.29 USD per operating hours. The other maintenance works can be estimated to be 0.13 USD/hr. Thus, the oil changing and maintenance cost can be totally calculated as 0.29 + 0.13 = 0.42 USD per operating hours.

The batteries have limited life time and must be replaced periodically, so we must include the price of batteries divided on the working hours to calculate the hourly operational costs. The cost of a 200Ah Yuasa lead acid battery is 300 USD [13]. The system contains 56 batteries; their total cost is 16'800 USD.

The following formula lets us evaluate the operating costs of the system under the management strategy:

$$P = \frac{(Pr_{Batteries} + Pr_{Diesel} \times V_{Diesel} + Pr_{maint} \times H_{Gen})}{H_{sys}} \qquad (12)$$

Where $Pr_{Batteries}$ is the total cost of batteries, Pr_{Diesel} is the cost of fuel per liter, V_{Diesel} is the consumed volume of diesel, Pr_{maint} is the cost of generator maintenance per operating hours, H_{Gen} is the generator operating hours and H_{sys} is the total operating hours of the system until batteries failure (Battery life time). In this way we can calculate the operating costs of the system per one hour.

The parameter we will work on optimizing is the minimum value of the state of charge reached by the battery storage system SOC_{min}. As it is expected, when the minimum allowed state of charge value is increased, this means that the dependence on the batteries decreases and the batteries life time will be extended. On the other hand the dependence on the diesel generators will increase, so the costs of diesel fuel and generator maintenance will increase also. This is shown in the Figure 12 and Figure 13.

The increase of battery life reduces the operation costs per hour, because the price of batteries will be divided on more working hours. The Figure 14 shows the results of simulation.

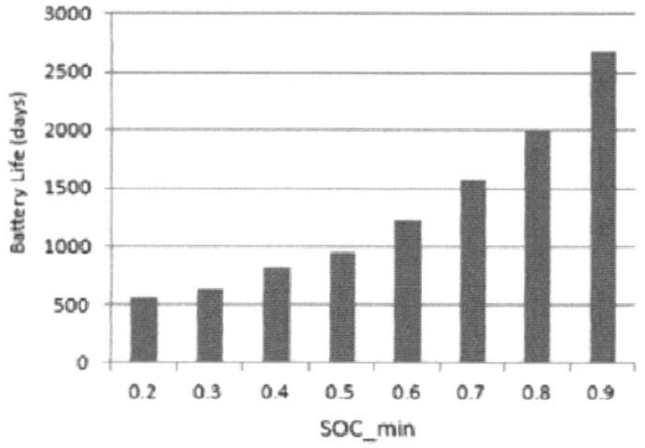

Figure 12: Battery Life as function of SOC_min

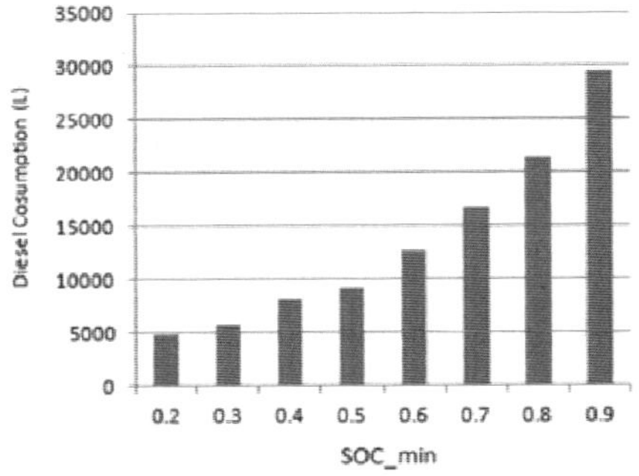

Figure 13: Annual fuel consumption as function of SOC_min

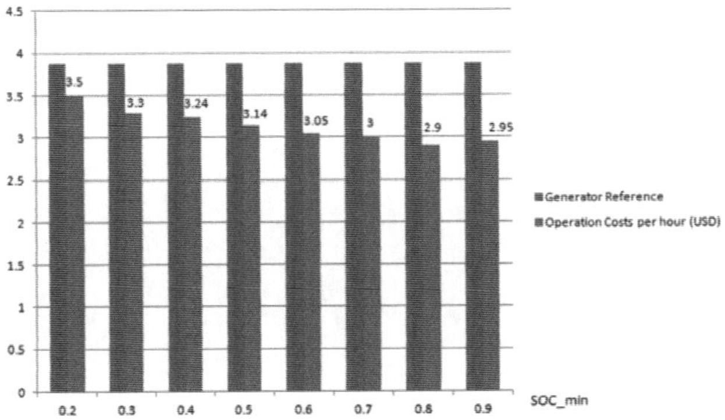

Figure 14: System operational costs as function of SOC_min

Figure 14 shows the operational costs of the system per hour as compared with the reference case, which is using only the diesel generator to run the system. It is obviously clear that all cases are less than the reference case, which confirms and validates more the strategy. Another result that we can see from this comparison is that at SOC_{min} = 0.8 we have minimum operation costs. This gives us a conclusion that it is not true to discharge the battery completely to get best advantage of it; there is a value between 0 and 1 that gives the best interest. Our calculated optimal SOC_{min} is 0.8, this value is subject to the real values of fuel price, battery price and maintenance costs.

VI. CONCLUSION

In this study, an advanced battery management strategy that controls the charging and discharging of the batteries in a hybrid renewable energy system was developed. The hybrid renewable energy system that we have chosen to incorporate our management strategy is a hybrid PV-Diesel-battery off-grid system. The system was explained in details in all its components starting from the PV solar system to battery storage system to the diesel generator reaching the smallest components. The components have been developed into mathematical models. The model of the PV system has given power, voltage and current output from the MPPT controller. The batteries have been associated a model that estimates the consumed charging current, calculates the change of state of charge upon charging or discharging of the battery, counts the consumed life time cycles of the battery and consider the evolution of the battery capacity after degradation with time. The diesel generator model has focused on counting the quantity of consumed diesel fuel during the operation of the system. Before developing the battery management strategy, we have determined the necessary and recommended conditions and constraints that must be respected in the strategy. A case study has been taken and its hybrid system has been sized, thus the management strategy can be simulated on a real base.

We have validated that the developed battery management strategy can control the power flow in a hybrid renewable energy system to charge and discharge the battery in the right way and the right time, taking into consideration the battery protection, life time and capacity degradation. As to optimize the strategy, we worked on optimizing some parameters. A good indication to evaluate the management strategy is the operational cost per hour of the system; this value was used in our comparative optimization. The first parameter that was optimized was the minimal allowed value of state of charge of the batteries. We run our system in same conditions and varied SOC_{min} ; the minimal operating cost was at $SOC_{min} = 0.8$. Another parameter can be studied and presented in another works such as the number of batteries in the system and the number of the diesel generator.

As a final conclusion, we can say that using an efficient battery management strategy in a hybrid renewable energy system can increase the performance of the system and make it more reliable, it can decrease the operational cost of the system by extending the life time of the batteries and can decrease the investment costs by decreasing the number of needed batteries for the same reliability and efficiency.

REFERENCES

[1] A. Woyte, V. Van Thong, R. Belman and J. Nijs, "Voltage fluctuations on distribution level introduced by photovoltaic systems," IEEE Trans. Energy Convers., vol. 21, no. 1, pp. 202-209, Mar. 2006.

[2] P. Denholma and R. Margolis, "Evaluating the limits of solar photovoltaics next term (PV) in traditional electric power systems," Energy Policy, vol. 35, p. 2852–2861, 2007.

[3] Y. Riffonneau, S. Bacha, F. Barruel and S. Ploix, "Optimal Power Flow Management for Grid Connected PV Systems With Batteries," IEEE TRANSACTIONS ON SUSTAINABLE ENERGY, vol. 2, no. 3, pp. 309-320, July 2011.

[4] P. Denholma and R. Margolis, "Evaluating the limits of solar photovoltaics (PV) in electric power systems utilizing energy storage and other enabling technologies," Energy Policy, vol. 35, p. 4424–4433, 2007.

[5] M. Perrin, Y. Saint-Drenan, F. Mattera and P. Malbranche, "Lead-acid batteries in stationary applications: Competitors and new markets for large penetration of renewable energies," Power Sources, vol. 144, p. 402–410, 2005.

[6] G. Coppez, Optimal Sizing of Hybrid Renewable Energy Systems for Rural Electrification, The University of Cape Town , 29 August 2011, p. The University of Cape Town.

[7] "Caterpillar - Diesel Generator Set," [Online]. Available: http://www.olympianpower.com/olympian-diesel-gensets/50hz.

[8] R. Dufo-López, J. Lujano-Rojas and J. Bernal-Agustín , "Comparison of different lead–acid battery lifetime prediction models for use in

simulation of stand-alone photovoltaic systems," Applied Energy , vol. 115, p. 242–253, 2014.

[9] B. Idlbi, Dynamic Simulation of a Pv-Diesel-Battery Hybrid Plant for Off Grid Electricity, Kassel, Germany: Faculty of Engineering, University of Kassel, March 2012.

[10] H. Ibrahim, Étude et conception d'un générateur hybride d'électricité de type éolien-diesel avec élément de stockage d'air comprimé, Chicoutimi, Canada: Thèse de doctorat. Université du Québec à Chicoutimi, 2010.

[11] H. Ibrahim, A. Ilinca and J. Perron, "Integration of wind turbines with compressed air energy storage in remote area power supply system," in Procceedings of European Wind Energy Association (EWEA), Copenhagen, Denmark, 2012.

[12] M. Torres and L. Lopes, "Inverter-Based Diesel Generator Emulator for the Study of Frequency Variations in a Laboratory-Scale Autonomous Power System," Energy and Power Engineering, vol. 5, no. 3, p. 10 , 2013.

[13] Diesel Engine - Combustion, Emissions and Condition Monitoring, Intech Open Science, April 30, 2013.

[14] H. Ibrahim , M. Dimitrova, I. A. and P. J., "Système hybride éolien-diesel avec stockage d'air comprimé pour l'électrification d'une station de télécommunications isolée," Revue internationale de génie électrique, vol. 12, no. 5-6, pp. 701-731, 2009.

[15] M.A. Islam, A. Merabet, R. Beguenane, H. Ibrahim. Power management strategy for solar-wind-diesel stand-alone hybrid energy system. World Academy of Science, Engineering and Technology, International Journal of Electrical, Computer, Energetic, Electronic and Communication Engineering, Vol. 8, N. 6, Pages 843-847, 01 May 2014.

[16] K Belmokhtar, H Ibrahim, Z Féger, M Ghandour. Charge Equalization Systems for Serial Valve Regulated Lead-Acid (VRLA) Connected Batteries in Hybrid Power Systems Applications. Energy Procedia, Vol. 99, Pages 277-284, 30 November 2016.

[17] M Rezkallah, Ambrish Chandra, DR Rousse, H Ibrahim, A Ilinca, D Ramdenee. Control of small-scale wind/diesel/battery hybrid standalone power generation system based on fixed speed generators for remote areas. Industrial Electronics Society, IECON 2016-42nd Annual Conference of the IEEE, Pages 4060-4065, 23-26 Oct. 2016, Florence, Italy.

[18] H Hosni, S Benhamed, H Ibrahim, M Ghandour, A Ilinca, K Belmokhtar. Power flow management strategy for renewable hybrid energy system. IEEE-Electrical Power and Energy Conference (EPEC), Pages 1-6, 12-14 Oct. 2016, Ottawa, Canada.

[19] K Belmokhtar, H Ibrahim, M Ghandour. Improving performance of batteries by using charge equalization systems—Experimental validation. IEEE-3rd International Conference on Renewable Energies for Developing Countries (REDEC), Pages 1-5, 13-15 July 2016, Beirut, Lebanon

A Novel Design of Water-Flow Based Electrical Generator as an Energy Harvesting Device

Hamdy A. Ziedan
Electrical Engineering Department,
Faculty of Engineering, Assiut University, Egypt
College of Engineering, AlJouf University, Saudi Arabia
E-mail: ziedan092@yahoo.com

Ibraheem M. Fayed
Network Planning Department, National Telecommunication Institute (NTI), Cairo, Egypt.
E-mail: imfayed@ju.edu.sa

Alaa Eldin M. Abofard
Alexandria Higher Institute of Engineering & Technology (AIET),
Alexandria, Egypt
E-mail: alaaabofard@yahoo.com

Abstract—This paper is aimed to investigate, experimentally, new two designs of Energy Harvesting Device: Water-Flow Based Electrical Generator (WEG). This device directly converts the motion of water inside a permanent magnet tube to electricity. Its output power is affected by several parameters such as permanent-magnet tube diameter, its length, water velocity and the concentration of minerals in the water (fresh, sewage or sea water). The relationship and dependency of these parameters are examined experimentally to obtain a set of graphs which can determine the suitability and performance of the proposed model for different applications. Advantages of these new models are clean power resource, simple and non-expensive design. The output power of these generators can power or recharge cell phones, mobile computers, radio communication equipment, stairs lighting of residential towers, commercial street signs, inside sea signs, etc.

Keywords—Energy harvesting devices; Water-Flow Based Electrical Generator (WEG); Permanent magnet tube; Water velocity.

I. INTRODUCTION

Energy harvesting materials and systems have emerged as a prominent research area and continue to grow at rapid pace. A wide range of applications are targeted for the harvesters, including distributed wireless sensor nodes for structural health monitoring, embedded and implanted sensor nodes for medical applications, recharging the batteries of large systems, monitoring tire pressure in automobiles, powering unmanned vehicles, and running security systems in household conditions. Recent development includes the components and devices at micro–macro scales covering materials, electronics, and integration. Energy harvesting devices converting ambient energy into electrical energy have attracted much interest in both the military and commercial sectors. Some systems convert motion, such as that of ocean waves, moving of human body, etc. [1-3].

Water is one of the most abundant elements on Earth. It is the only item to exist in all three states of matter at once naturally, but what makes water truly unique are its "individual properties." These features include its high surface tension and cohesion, neutral PH, high electrical resistance, high specific heat, and capacity as the "universal solvent" [4]. There are several studies discussed the effects of several variables on properties of water itself. Banejad and Abdo Saleh examined the relationship between magnetic fields and flow rate in diamagnetic fluids, [5]. While the effect of the magnetic field intensities of zero Tesla, 0.05 Tesla, 0.075 Tesla, and 0.1 Tesla, on the water hardness were examined, [5]. The electromotive force generation with Hydrogen release by salt water flow under a transverse magnetic field was studied by Roberto De Luca, [6]. Michael Faraday found experimentally in 1831 that a time-varying magnetic field gives rise to an electric field which is known by "Faraday's law," [7]. The effect of concentration of minerals on water as a conducting material was studied by Ziedan et al. in [8].

This paper is aimed to investigate a novel two designs of energy harvesting device to generate electrical power from natural/forced flow of water inside a permanent magnetic tube, which called Water-Flow based Electrical Generator (WEG). This type of generators has no hydraulic turbine. It depends on Faraday's law. Two new models of WEG are designed and tested experimentally. The output power of WEG depends on water velocity, length, the diameter of the permanent magnetic tube, and concentration of minerals in the water. All these variables are studied carefully in experimental work.

II. METHODOLOGY

The electromotive force around a closed path C is equal to negative of the time rate of increase the magnetic flux enclosed by that close road [7], as shown in the following equation:

$$\oint_c E \,.dl = -\frac{d}{dt} DDT \int_{s} B \,.\, ds \tag{1}$$

Where E is the electric field strength in (V/m), and B is the magnetic flux density in Tesla (T). This can be represented by the first Maxwell equation as shown in (2):

$$\nabla \times E = -\frac{\partial B}{\partial t} \tag{2}$$

Permanent magnet tube produces a magnetic flux in radial directions cuts water in normal angle, Figure 1-5. Water has good electrical conducting properties depends on salt concentrations [4, 8]. Water velocity v was used to control time-varying in the magnetic field of a permanent magnet tube to generate electromotive force around a closed path through the magnetic tube. A non-magnetic grid placed at one terminal of the magnetic tube, the end side of permanent magnet tube in the first model only, Figure 2. This non-magnetic grid is used to transfer electricity from water to load circuit. Time-varying of the magnetic field will be controlled by water velocity v through a tube cross-sectional area S is expressed as shown in the following equation [9]:

$$Q = S \times v \tag{3}$$

Where; Q is flow rate or discharge in (m^3/sec), S is a cross-sectional area in (m^2), and v is water velocity in (m/s).

III. Experimental Set-Up

Water-Flow based Electrical Generator (WEG) was set-up in Electrical Power Laboratory, AlJouf University, Saudi Arabia. The set-up made it possible to measure the effect of varying water velocity v, length L, diameter D of the permanent magnetic tube, the concentration of water minerals with an output power of WEG units. There are two new models of WEG are presented in this paper.

A. *First new model of WEG*

Description of this type of WEG is shown in the following points:

- A tube of permanent magnet material of Cast Alnico magnet with 25, 50, 75 and 100 cm length L, inner diameter D of the tube is 2.5, and 5 cm, thickness of tube t is 0.5 cm, Figures 1, 2 and 3.
- Water flows through a permanent magnet tube with velocity v of 1, 2, 3 and 4 m/sec.
- None-magnetic cylindrical grid, Figures 2 and 3, is used to transfer electricity from water to load circuit.
- The electric pump is used to control water velocity v, Figure 3.
- Tank to store water, Figure 3.

B. *Second new model of WEG:*

In this type of WEG, a PVC tube of 1 m length is used. Copper wire of radii 0.5 mm is surrounded a PVC tube.

IV. Experimental Technique:

In the 1st model of WEG, water is flowing through a magnetic tube by certain velocity v controlled by an electric pump. None-magnetic cylindrical grid, Figures 2 and 3, is used to transfer electricity from water to load circuit. Induced emf is measured between non-magnetic cylinder and ground terminal by a digital voltmeter. Water velocity v is controlled by a valve placed close to the electric pump.

In the 2nd model of WEG, induced emf is measured between two copper wire terminals by a digital voltmeter, Figures 4 and 5.

Over the copper wire, a strong magnetic cylinder is placed. Induced emf is measured between copper wire terminals as shown in Figures 4 and 5.

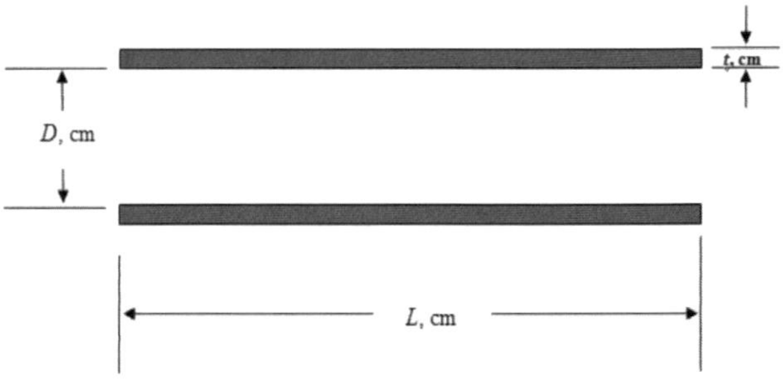

Fig. 1. Cylindrical permanent magnetic tube from Cast Alnico magnet.

V. Results and Discussion

Induced emf of the 1st model of WEG is increased by increasing water velocity v for same inner tube diameter and same tube length, Figure 6 - 7. Also, as inner permanent magnetic tube diameter increased the induced emf increased too, Figure 8 - 11. And it is increased with the increasing of the length of permanent magnetic tube for same water velocity v and same inner tube diameter, Figures 6 – 11.

Figure 12 shows the effect of concentration of minerals of water (fresh, sewage or sea water) on induced emf of 1st model WEG unit. Induced emf is increased by increasing of minerals in water due to increase water conductivity as reported in [8].

Induced emf of the 2nd model of WEG is affected by water velocity v, permanent magnetic tube diameter D, and length L as the 1st model except for the magnitude of induced emf is lower than in 1st model, Figure 13.

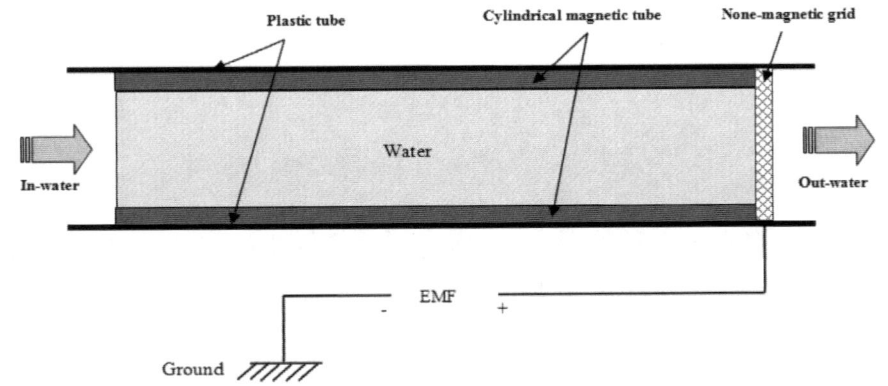

Fig. 2. Schematic diagram of the 1st model of Water-Flow based Electrical Generator (WEG) Unit.

Fig. 3. Prototype model of the 1st model of WEG Unit (AlJouf University).

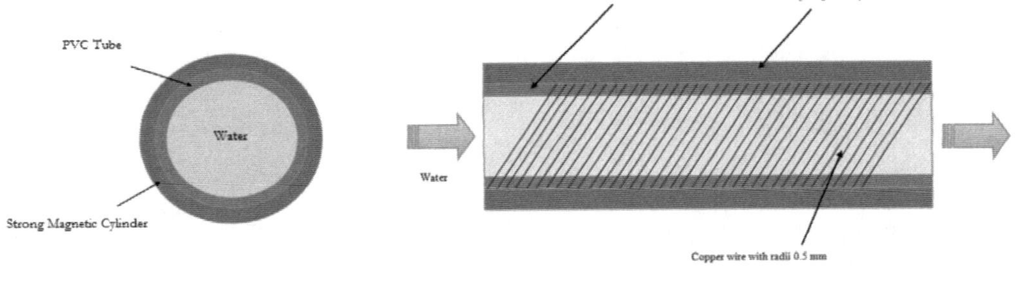

Fig. 4. Schematic diagram of the 2nd new design of WEG unit.

https://doi.org/10.1051/matecconf/201817102001

Fig. 5. Prototype model of the 2nd model of WEG Unit (AlJouf University).

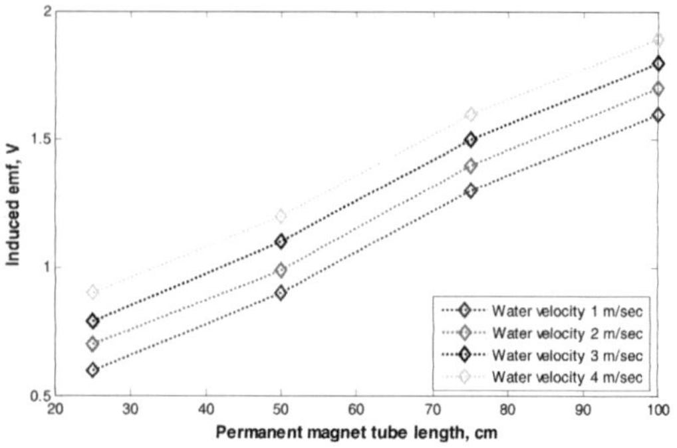

Fig. 6. Effect of varying tube length with different water velocity, v on induced emf of 1st model WEG unit. (Inner tube diameter, D = 2.5 cm)

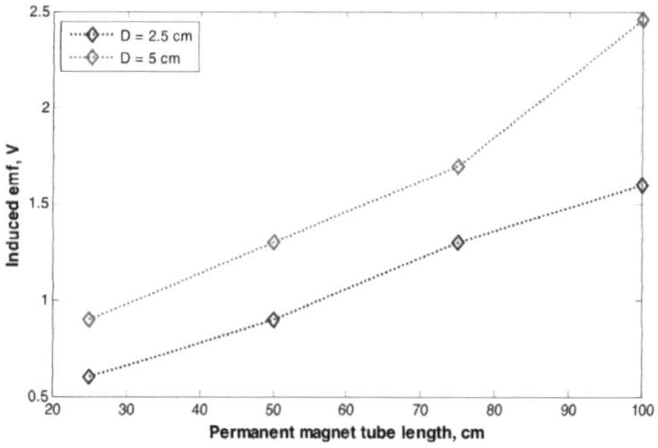

Fig. 8. Effect of varying inner tube diameter D on induced emf of 1st model WEG unit, (Water velocity, v = 1 m/sec).

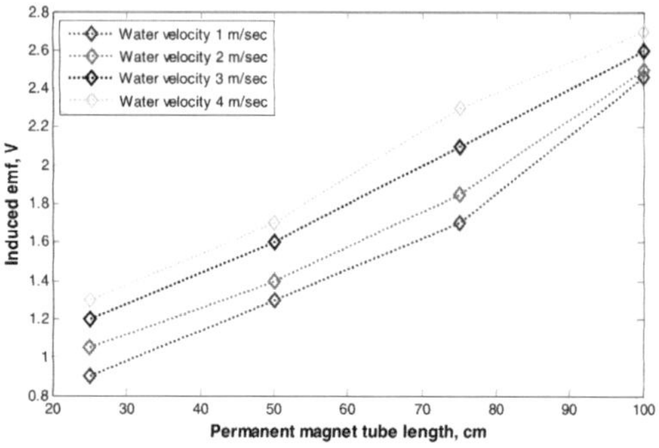

Fig. 7. Effect of varying tube length with different water velocity, v on induced emf of 1st model WEG unit, (Inner tube diameter, D = 5 cm).

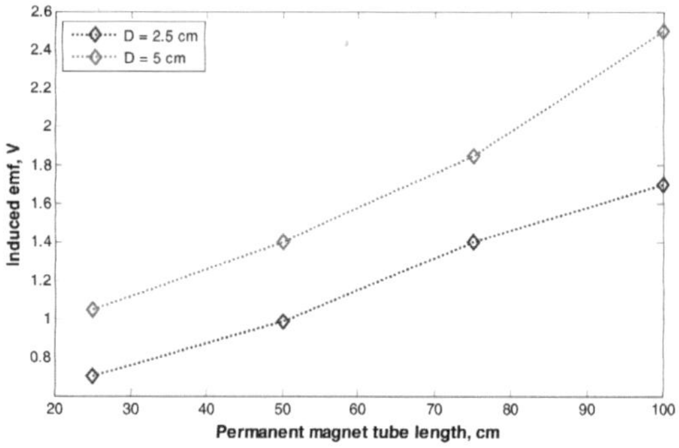

Fig. 9. Effect of varying inner tube diameter D on induced emf of 1st model WEG unit, (Water velocity, v = 2 m/sec).

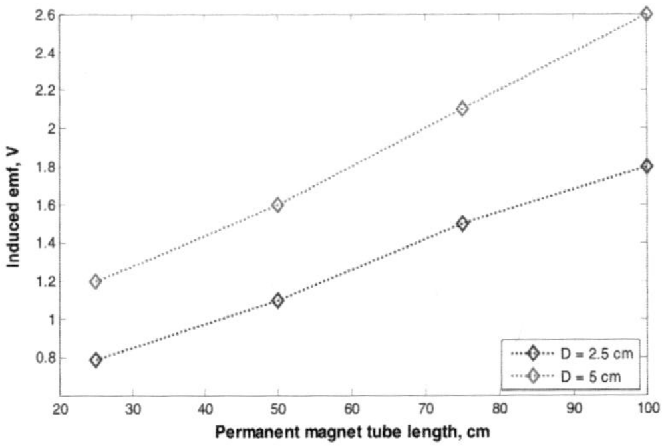

Fig. 10. Effect of varying inner tube diameter D on induced emf of 1st model WEG unit, (Water velocity, v = 3 m/sec).

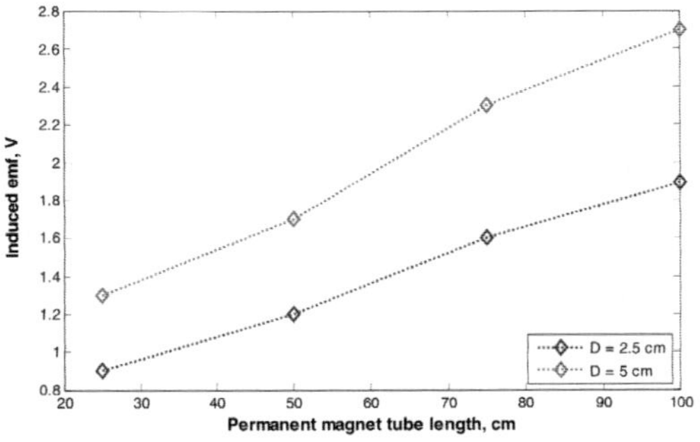

Fig. 11. Effect of varying inner tube diameter D on induced emf of 1st model WEG unit, (Water velocity, v = 4 m/sec).

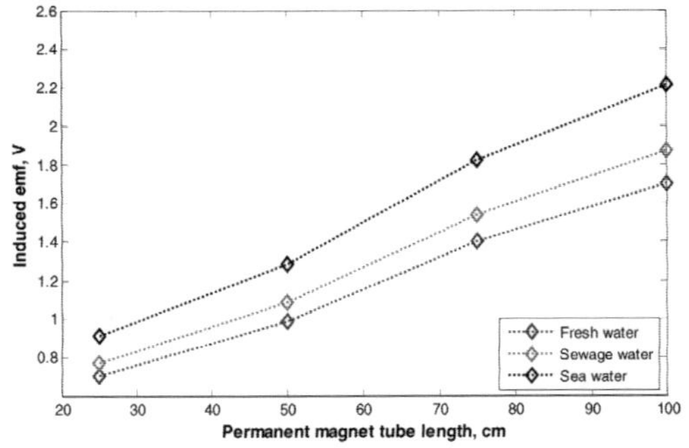

Fig. 12. Effect of concentration of minerals of water (fresh, sewage or sea water) on induced emf of 1st model WEG unit. (Inner tube diameter, D = 5 cm and Water velocity, v = 2 m/sec).

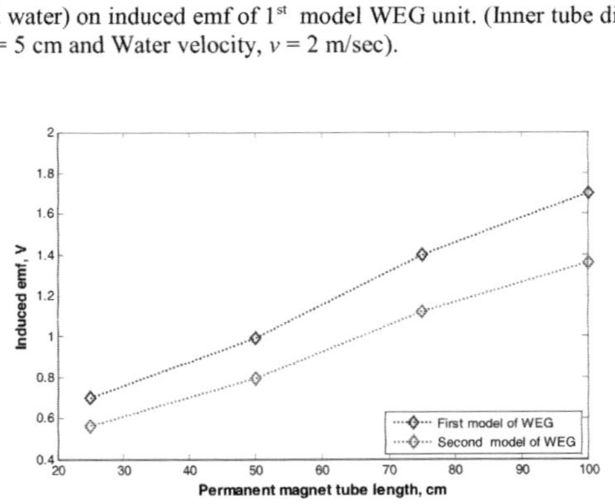

Fig. 13. Induced emf from 1st and 2nd models of WEG, (Inner tube diameter, D = 5 cm and Water velocity, v = 2 m/sec).

VI. Conclusion

Novel two models of Water-Electrical-Generator (WEG) as an Energy Harvesting Device are designed and tested experimentally. WEG depends on natural water flow through a permanent magnet tube. The first model is more power efficient than the second one. WEG models are used to generate electrical power; its rate depends on the dimensions of the system; length, diameter, and thickness of permanent magnet tube flow velocity of water inside this tube, the concentration of minerals in the water (fresh, sewage or sea water). This electrical power is a clean and not expensive power. The amount of generated electrical power can be controlled by several parameters such as water flow rate, magnetic tube length and inner magnetic tube diameter. This electrical power can be used in several personal applications such as Stairs lighting between floors depend on the natural flow of water from water tanks to supply apartments. Commercial street signs can be energized depend on water flow of fresh water or sewage. By using rechargeable batteries, the electrical power generated during water flow may be stored and be used through any moving application, cell phones, mobile computers, radio communication equipment, inside sea signs, etc.

VII. References

[1] Shashank Priya and Daniel J. Inman, "Energy Harvesting Technologies", Book, Springer Science and Business Media, 2009.

[2] S. Chalasani and J. M. Conrad, "A Survey of Energy Harvesting Sources for Embedded Systems", *Southeastcon IEEE*, pp. 442-447, 2008.

[3] M. El-Hami, P. Glynnes-Jones, N. M. White, M. Hill, S. Beeby, E. James, A. D. Brown and J. N. Ross, "Design and fabrication of a new vibration-based electromechanical power generator", *Sen. Actuators A*, 92, pp.335-342, 2001.

[4] "Water Properties: Facts and Figures About Water" USGS Water Science School, 13 Sept. 2012. http://ga.water.usgs.gov/edu/waterproperties.html.

[5] H. Banejad and E. Abdosalehi, "The effect of magnetic field on water hardness reducing", Thirteenth International Water Technology Conference, IWTC 13, Hurghada, Egypt, pp. 117-128, 2009.

[6] Roberto De Luca, "Electromotive Force Generation with Hydrogen Release by Salt Water Flow under a Transverse Magnetic Field", Journal of Modern Physics (JMP), 2, pp.1115-1119, 2011.

[7] Rajeev Bansal, "Fundamentals of Engineering Electromagnetics", Taylor & Francis Group, 2006.

[8] Hamdy Ziedan, Hesham Aljeelani, Ahmed Alanazi and Ibrahim Alshammari, "Electrical Characteristics of Holy Zamzam Water", 16th Scientific Symposium for Hajj, Umrah & Madinah Visit Scientific Portal for 1437AH, The Custodian of the Tow Holy Mosques Institute for Research of Hajj and Umrah, Makkah, Saudi Arabia, 24 - 25 May 2016.

[9] Zane Satterfield, P. E., "Fundamentals of Hydraulics Flow", Tech Brief, Vol. 10, Issue 1, pp. 1-4, Spring/Summer 2010.

VIII. ACKNOWLEDGMENT

The authors would like to thank *Prof. Ali M. Eltamaly*, of Mansoura University, Egypt and *Dr. Tarek Kandil*, of Al-Azhar University, Egypt for their interest in this research work.

IX. AUTHORS BIOGRAPHIC

Hamdy A. ZIEDAN, Associate Professor, was born in New-Valley, Egypt, on 1981. He received the B.Sc. and M.Sc. degrees from Minia University, Egypt, in 2003 and 2006 respectively. He was a Ph.D. student at High Voltage Laboratory, Faculty of Electrical Engineering, Czech Technical University in Prague, the Czech Republic from 2009 to 2010. He received his Ph.D. degree in High Voltage Engineering from Faculty of Engineering, Assiut University, Egypt and Czech Technical University in Prague, the Czech Republic in 2011, according to a Governmental Scholarship Program, which means a Scientific Cooperation between the Czech Technical University in Prague, Czech Republic and Assiut University, Egypt. He became an Associate Professor in Electrical Power Engineering from July 2016. Since 2004, he has been with the Department of Electrical Engineering, Faculty of Engineering, Minia University and Assiut University as a Teaching Assistant, Lecturer Assistant, and since 2011, as an Assistant Professor. He was a Visiting Researcher at Czech Technical University in Prague, Czech Republic, from 2012 to 2013. He is an Assistant Professor at Electrical Engineering Department, Faculty of Engineering, AlJouf University, Saudi Arabia from 2014 until now. His research interests include High Voltage Engineering, Electric Power Systems, and Renewable Energy.

Ibraheem M. Fayed, Assistant professor, B.Sc. in Electronics and Communication Engineering, (1995 Military Technical College), M. Sc. in Electronics and Communication Engineering, (2004 Arab Academy for Science and Technology), Ph.D. In Electronics and Communication Engineering, (Faculty of Engineering, Ain Shams University, 2010). He is working as an assistant professor, Networks Planning Dept. (NPD), National Telecommunication Institute (NTI), Ministry of Communication and Information Technology (MCIT), Egypt, which involves conducting both of postgraduate (Diploma) and NPD Engineering courses and conducting researches in computer, communication, and wireless networks since 1998 till Now.

Alaa Eldin M. Abofard, Assistant Professor, was born in Gharbia, Egypt, on 1969. He received the B.Sc., M.Sc and. Ph.D. degrees from Alexandria University, Egypt, in 1992, 2007 and 2011 respectively. He is an Academic Instructor at ADC (Air Defense College) and Academic Instructor visitor at Higher Institute for Engineering & Technology in Alexandria from July 2013 till now. From 2008 to 2012, he was an Academic Assistant Instructor at ADC (Air Defense College) and Academic Assistant Instructor visitor at Higher Institute of Engineering & Technology in Behira, Egypt. From 2001 to 2017, he was an Academic Assistant Instructor at ADC (Air Defense College). His research interests include Electric Power Systems, Renewable Energy, and High Voltage Engineering.

 https://doi.org/10.1051/matecconf/201817102002

Design of a Hybrid Photovoltaic Thermal System in Lebanon

Elie Karam, Patrick Moukarzel, Maya Chamoun, Charbel Habchi and Charbel Bou-Mosleh*
Mechanical Engineering Department
Faculty of Engineering, Notre Dame University-Louaize
Zouk Mosbeh, Lebanon
* Corresponding author: cboumosleh@ndu.edu.lb

Abstract—**Due to global warming and the high toxic gas emissions of traditional power generation methods, renewable energy has become a very active topic in many applications. This study focuses on one versatile type of solar energy: Hybrid Photovoltaic Thermal System (hybrid PV/T). Hybrid PV/T combines both PV and thermal application and by doing this the efficiency of the system will increase by taking advantage of the temperature loss from PV module. The solar radiation and heat will be harnessed to deliver electricity and hot water simultaneously. In the present study a solar system is designed to recycle the heat and improve the temperature loss from PV module in order to supply both electricity and domestic hot water. The project was tested twice in Zouk Mosbeh - Lebanon; on May 18, 2016, and June 7, 2016. The average electrical efficiency was around 11.5% with an average electrical power output of 174.22 W, while with cooling, the average electrical efficiency reaches 11% with a power output of 200 W. The temperature increases by about 7 degrees Celsius from the inlet. The 1D conduction model is also performed in order to design the hybrid PV/T system.**

Keywords—renewable energy; hybrid; photovoltaic; heat transfer; efficiency.

I. INTRODUCTION

There are two types of solar panels that collect Sun's light and heat: solar water heating panels and photovoltaic solar panels Solar water heaters, also known as solar domestic hot water systems, use solar radiation in order to heat water for domestic or residential applications. Photovoltaic Solar Panels convert sunlight into electricity. The main desire of the researchers was to develop cells that have a low cost and a high efficiency [1]. The problem with photovoltaic panels is that their efficiency drops with temperature increase at a rate about 0.4-0.5% per 1°C once the baseline temperature of 25°C has been exceeded [2]. Several methods have been developed in order to limit the efficiency drop of PV panels by cooling them.

Recent study shows that the solar panel can attain a gain of 8 to 9% of net energy by pumping water over the surface of the panel. This continuous flow of water acts as a cooling method for the cells and a cleaning agent that reduces panel soiling [2].

Bojanampati *et al.* [3] experimentally investigated both air and water forced cooling methods. Their work made use of a Plexiglas U-channel which was in direct contact with the PV modules so as to remove thermal contact resistance. At an illuminance of 22 flux, the peak power output increase was 10% from passive to forced air cooling, 48% when water at 21°C was used, and 66% when chilled water at 14°C was used. Mehrotra *et al.* [4] investigated the effects of PV module emersion in water on its efficiency. It was reported that the efficiency of the panel increased with immersion depth up to a certain point, after which the efficiency began to decline. The maximum efficiency of 4.76 % was obtained under 1cm depth of water with the proposed design and operating condition.

The objective of this study is to design a solar system that recycles the heat and improves the temperature loss from PV module in order to supply electricity and domestic hot water.

II. PROTOTYPE IMPLEMENTATION

The copper tube piping network design was chosen to be a parallel pipe system rather than a serpentine pipe system as shown in Figure 1. The parallel riser tubes are designed to transport collector fluid from the bottom of the collector to the top. The headers have a larger diameter than the risers because fluid pressure is greatest at the base at the incoming fluid and least by the output pipe, so the pressure difference is moderate and the flowrate in each of the parallel pipes is more uniform. The flowrate is minimal at the center of the collector due to its smaller diameter where most of the heat is concentrated. An insulation cover is used on the back of the panel to reduce the heat losses to ambient air.

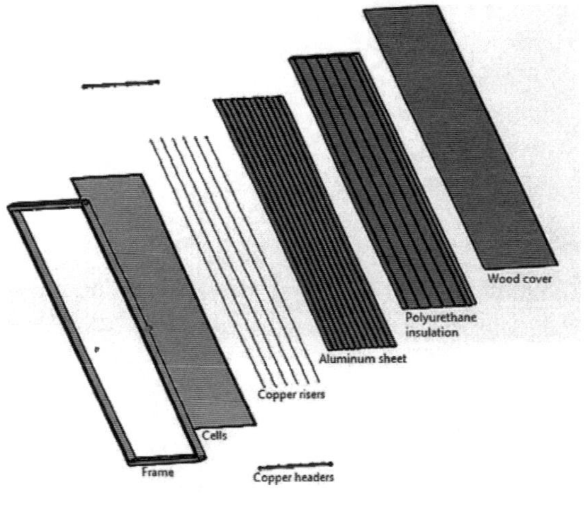

Figure 1: The different parts of the PV/T system

 https://doi.org/10.1051/matecconf/201817102002

The pump is selected based on the head and the flow rate. The flow rate is assumed to be equal to 0.8 GPM and the head should be equal to the head loss of the system. The total head loss is equal to the summation of minor and major losses due to the fittings and the pipe equal to 4 cm.

The 1D thermal resistance network model is used to determine a convenient polyurethane insulation thickness for the back of the PV/T as shown in Figure 2. First the heat is lost from the water to the inner riser walls by convection. Then the heat loss is conducted to the ambient air through several conducting media [5]. Two assumptions are made here: the water average bulk temperature is 40°C and the ambient air temperature is 25°C. The calculation shows that an insulation thickness of 6.5 mm is needed for the prototype.

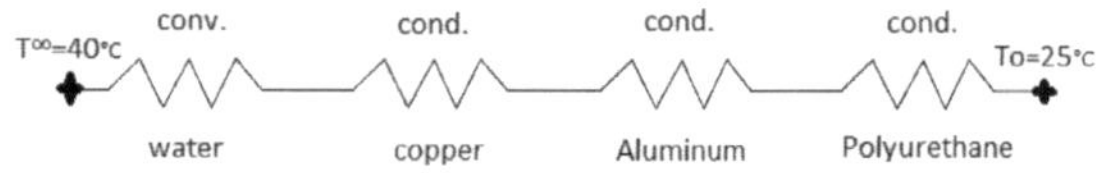

Figure 2: Thermal resistance network

III. Experimental Study

The system was tested twice, the first time on Wednesday May 18, 2016, and then on June 6 and 7 2016. Figure 3 shows a schematic diagram of the system. The setup shown in Figure 4 consists of the hybrid PV/T panel, insulated PEX pipes, a 60 L tank, 12V DC pump and 12V battery, a pyranometer to measure the solar irradiance, flowmeter, voltmeter, digital thermometer, thermocouple, six 12V LED lamps each rated at 50W, and a multimeter. The panel was oriented with an inclination of 33° to the south due to the solar positioning consideration.

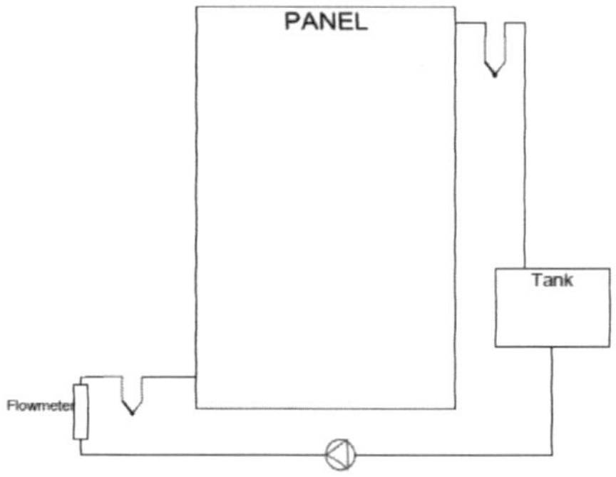

Figure 3: Schematic diagram of the system

Figure 4: The experimental setup

Figure 5 shows the panel and fluid temporal temperature variation. It is observed that the panel temperature is around 49°C in May and around 56°C in June due to higher ambient air temperature in the later. The temperature difference in May is better than that obtained in June due to higher solar irradiance as shown later. In May the average temperature difference is around 5.4°C with a maximum of 8°C. The temperature difference in June drops to 3.3°C with a maximum around 5°C.

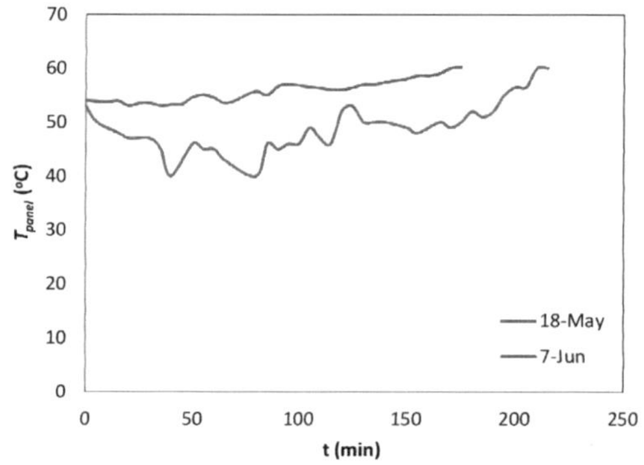

(a)

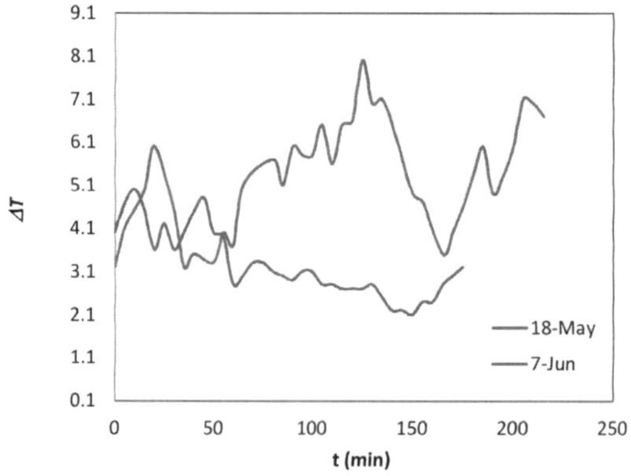

(b)

Figure 5: Temporal variation of the (a) panel temperature and (b) temperature difference between the inlet and the outlet

The thermal energy can be evaluated using the following equation:

$$E_{thermal} = \dot{m}\, c_p\, \Delta T \quad (1)$$

where ΔT is the temperature increase from the inlet to the outlet.

Knowing the dimensions of the panel the input power could be calculated as follows:

$$P_{in} = G \times A \quad (2)$$

with G the solar irradiance in W/m^2, and the area of the panel is $A = 1.626$ m^2.

The output power is obtained as follows:

$$P_{out} = V \times I \quad (3)$$

where V and I are the measured voltage and current output, respectively.

The electrical efficiency can be evaluated using the following equation.

$$\eta_{ele} = \frac{P_{out}}{P_{in}} \quad (4)$$

The thermal efficiency can be evaluated using the following equation.

$$\eta_{th} = \frac{E_{thermal}}{P_{in}} \quad (5)$$

Figure 6 shows the temporal variation of the thermal energy and input and output power. Figure 7 shows the temporal variation of the thermal and electrical efficiency. The average input power in May is around 1720 W while in June it drops to 1606 W with a decrease of around 6.6% explaining the results presented in the previous paragraph. For the output electrical power, the average is around 200 W for the experiment done in May while in June it drops to 182 W. It should be noted that the oscillations in the results were due the passing clouds which blocked the sun thus decreasing the irradiance. The average thermal power output evaluated using equation (1) is around 1408 W in May and 861 W for June. This is caused by the low temperature difference obtained in June relative to that obtained in May as shown earlier.

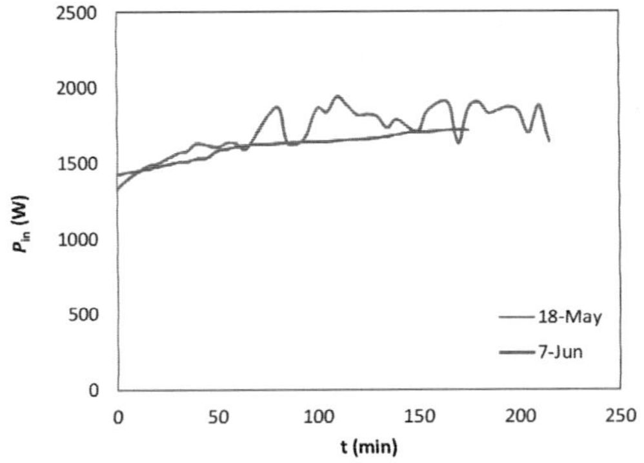

(a)

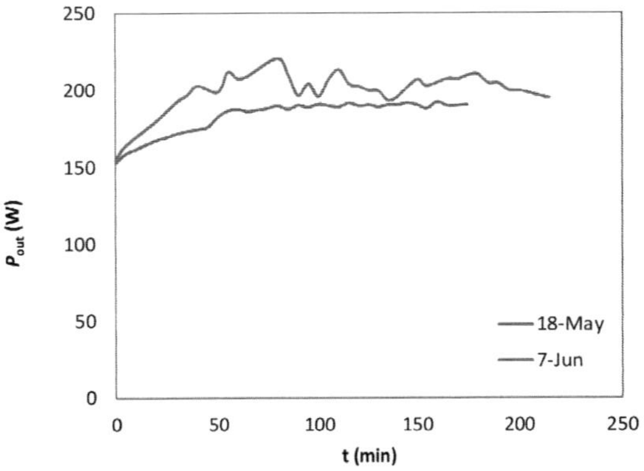

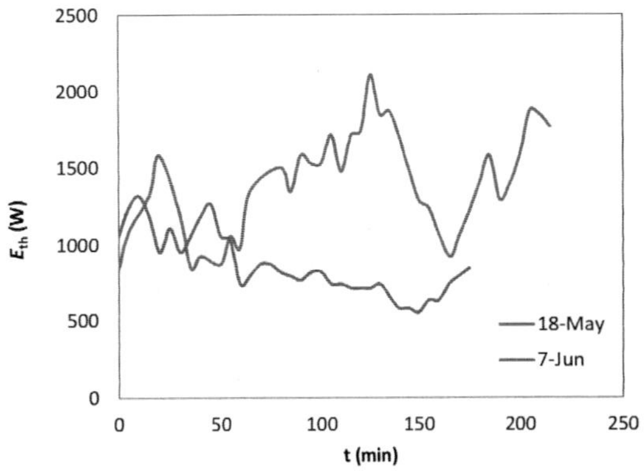

(c)

Figure 6: Temporal variation of (a) input power, (b) output electrical power and (c) thermal power

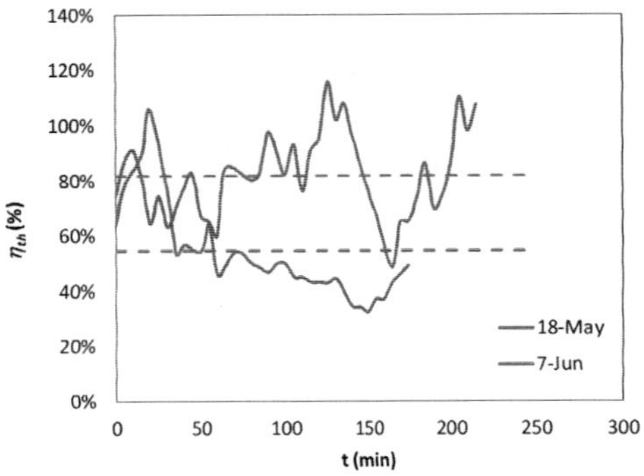

(a)

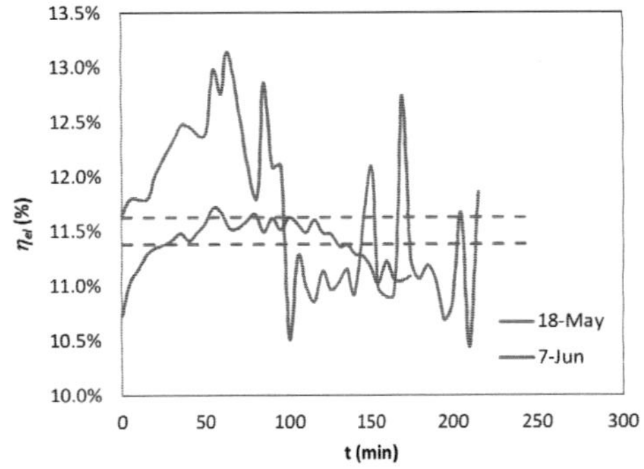

(b)

Figure 7: Temporal variation of the (a) thermal and (b) electrical efficiency

Figure 8 shows the variation of electrical efficiency in June for the case with and without cooling. As it can be observed from this figure, the cooling process enhances the efficiency of the PV panel by about 0.5%.

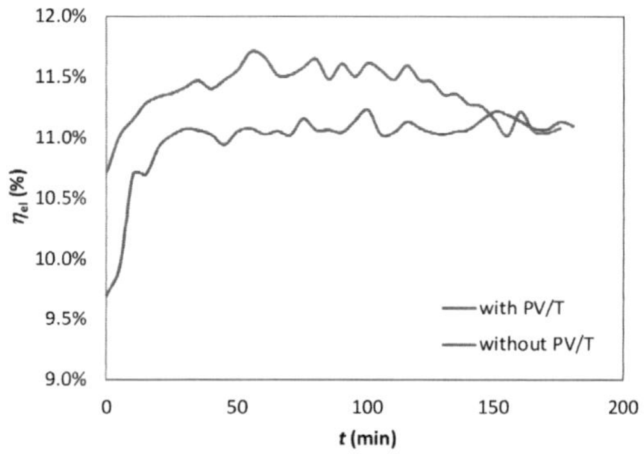

Figure 8: Temporal variation of the electrical efficiency in June for the cases with and without cooling

IV. Conclusion

The main objective of this study was to design and test a method to cool the photovoltaic cells with circulating water, and also take advantage of the heated water for domestic use. Hence, the efficiency of the panel will increase, and the circulating water will be hot enough for domestic use. The project was tested twice; on May 18, 2016, and June 7, 2016 in Lebanon.

An average of 11.5% electrical efficiency was achieved and an 82% average thermal efficiency on the first testing and 54% on the second one. The temperature difference between the inlet and the outlet reaches around 5.4°C for May and 3.3°C in June. The average electrical power calculated was around 200 W. The second testing was done to be able to compare the electrical efficiency with and without cooling. The average electrical efficiency without cooling was 11%, while an average of 11.5% with was achieved with cooling.

The electrical efficiency of the system did not increase as much as it was intended since the panel temperature did not reach the baseline of 25°C, but water was heated enough to be used in domestic applications. If the panel temperature was to be dropped to reach almost 25°C, chilled water could be used, but this will take more time to heat it.

Therefore, Hybrid PV/T harnesses the solar radiation and combines both PV and thermal application to deliver electricity and hot water simultaneously.

Finally, the results obtained so far in this project should encourage local and global sectors to think about alternative green power generation methods in order to decrease global warming and the high toxic gas emissions of traditional power generation.

References

[1] J. Andrews and N. Jelley, Energy sience: principles, technologies, and impacts, Oxford University Press, United Kingdom, 2013.

[2] M. K. Smith, H. Selbak, C. C. Wamser, N. U. Day, M. Krieske, D.J. Sailor, and T.N. Rosenstiel, "Water cooling method to improve the performance of field-mounted, insulated, and concentrating photovoltaic modules," Journal of Solar Energy Engineering. Vol. 136(3), pp. 034503-1-4, January 2014. doi: 10.1115/1.4026466.

[3] S. Bojanampati, P. Rodgers, and V. Eveloy, "Experimental assessment of flat-type photovoltaic module thermal behavior," Proeedings of the 13th International Conference on Thermal, Mechanical and Multi-Physics Simulation and Experiments in Microelectronics and Microsystems (EuroSimE), , April 16-18, 2012. doi: 10.1109/ESimE.2012.6191695.

[4] S. Mehrotra, P. Rawat, M. Debbarma, and K. Sudhakar, "Performance of a solar panel with water immersion cooling technique," International Journal of Science, Environment and Technology, Vol. (3), pp. 1161-1172, June 2014.

[5] J. H. Lienhard IV and J. H.Lienhard, A heat transfer textbook, Phlogiston Press, Cambridge, Massachusetts, USA, 2008.

Cooling Performance of Heat Sinks Used in Electronic Devices

Ibrahim Mjallal, Hussein Farhat, Mohammad Hammoud*, Samer Ali, Ali AL Shaer and Ali Assi
Mechanical an Electrical Departments, School of Engineering, Lebanese International University
Beirut, Lebanon
*Correspondence: mohamad.hammoud@liu.edu.lb

Abstract
Existing passive cooling solutions limit the short-term thermal output of systems, thereby either limiting instantaneous performance or requiring active cooling solutions. As the temperature of the electronic devices increases, their failure rate increases. That's why electrical devices should be cooled. Conventional electronic cooling systems usually consist of a metal heat sink coupled to a fan. This paper compares the heat distribution on a heat sink relative to different heat fluxes produced by electronic chips. The benefit of adding a fan is also investigated when high levels of heat generation are expected.

Keywords: Heat sink; Thermal management; Electronic devices

I. INTRODUCTION

Electronic equipment have made their way into practically every aspect of modern life, from toys and appliances to high-power computers. Recent increase in processing demands has forced manufacturers to increase the performance and functionality of integrated circuit chips in addition to minimizing their size leading to high power dissipation through smaller packages. This fact made thermal management of integrated circuit chips a critical aspect of successful processor system design, and finding new cooling techniques much more valuable. Moreover, the failure rate of electronic equipments increases exponentially with temperature. Higher temperatures, increases power leakage and degrades the chip reliability. Generally, the global maximum allowable temperatures of various chips range between 85°C and 120 °C [1]. To avoid such problem, packages of integrated circuits must be designed to remove any predicted excessive heat.

Since high temperature reduces performance and degrades chip reliability, the effective management of processor's temperature is an important issue that needs to be investigated. In order to avoid excessive levels of heat and reduce temperature, active and passive cooling techniques are used. In many applications, standard cooling methods could be insufficient for the heat load produced during continuous operation over long periods due to some limitations such as limited thermal conductivity of air for convection and copper for conduction and small spaces decreasing the capabilities performance.

In the present work, thermal management of the chip's temperature is performed using an aluminum heat sink that can dissipate the heat generated by the chip and release it to the surrounding.

II. CLASSIC COOLING TECHNIQUES

For years, various active and passive cooling methods have been applied to cool electronic chips. Active methods are those who need power to work such as forced ventilation using electric fans, liquid cooling through microchannels [2] and heat pipes in fanless systems [3]. While in the Passive cooling methods no power is needed to initiate the cooling process as well as no mechanical moving parts are used. Heat sinks is an attractive option for low power systems. However, size constraints make the use of heat sinks difficult in embedded or mobile systems.

A. Heat Sink

A heat sink is a passive heat exchanger that transfers the heat generated by an electronic device (chip) to a lower temperature fluid medium [4]. The main role of the heat sink is to maximizes the surface area of contact between the processor and the surrounding cooling medium, by the use of fins. Heat generated by the chip is transferred to the heat sink by conduction, Subsequently this heat is transferred to the surrounding, which is usually air, by convection and radiation. Most often, heat spreaders are used between the chip and the heat sink due to the difference in size between the bottom of the sink and the top of the chip, to ensure more heat dissipation. A heat spreader is a heat exchanger that moves heat between a heat source and another secondary heat exchanger whose surface area and geometry are more favorable than the source. Such a spreader is most often simply a plate made of copper, which has a high thermal conductivity.

Heat sinks as shown in Figure 1 are usually made of metals such as aluminum or copper both having high thermal conductivity to minimize the temperature difference between the chip and the fin's tip.

The cooling package is modeled as a thermal resistance circuit on basis of R (thermal resistance) or C (thermal conductance) where the temperature across two locations can be measured, leading to the calculation of heat dissipation rate [5].

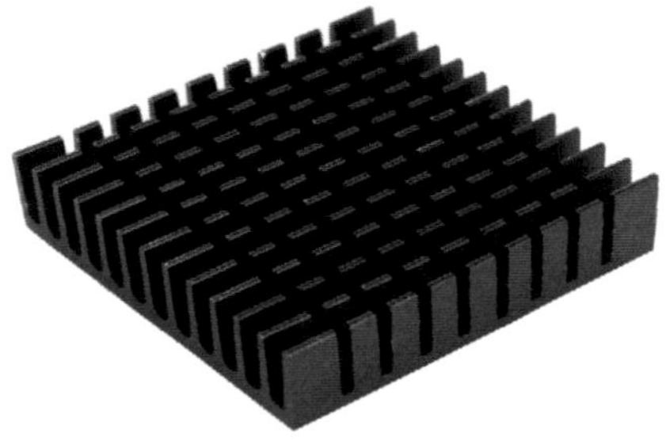

Fig.1. Heat Sink [6]

B. *Other Cooling Methods*

Various active and passive cooling methods have been applied to cool electronic chips. Such as forced ventilation, heat pipes, liquid cooled cold plates, immersion cooling, and the use of Phase Change Materials [7].

III. EXPEREMENTAL SETUP

The experimental setup is shown schematically in Fig. 2. The heat sink is placed on top of an electronic chip. The dimensions of the sink are 20×20×6 mm. Both the chip and the heat sink are drawn using SolidWorks in order to be used in further simulation.

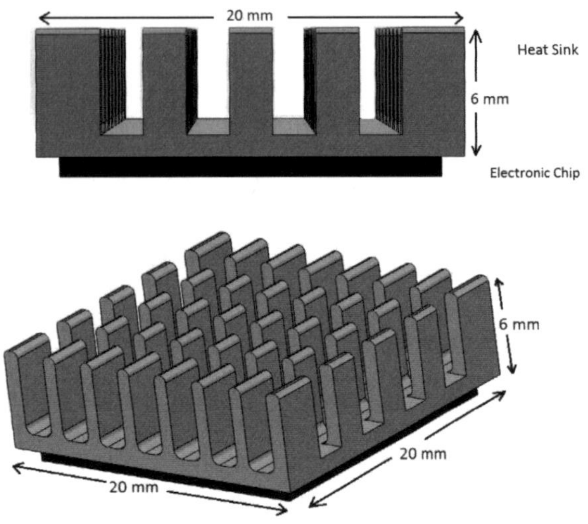

Fig. 2. SolidWorks drawing of the simulated domain.

A. *Heat Transfer*

The heat generated by the electronic chip is transferred through the heat sink by conduction and then the heat is carried out by the surrounding air by convection (natural convection is considered in our study) and radiation. Usually, a heat spreader is used to ensure the spreading of the heat all over the whole base of the sink and to overcome the size difference between the electronic chip and the heat sink.

B. *Equations*

1) Conduction: is the transfer of energy from the more energetic, to less energetic particles of the same substance or between two substances in contact due to interaction between particles. [6]

$$q" = -K.\nabla T \quad (1)$$

Where;

$q"$ is the heat transfer rate (W/m^2)

K is the thermal conductivity $(W/m.^{\circ}K)$

$$\nabla T = \left(i\frac{\partial T}{\partial x} + j\frac{\partial T}{\partial y} + k\frac{\partial T}{\partial z} \right) \quad (2)$$

2) Convection: occurs between a fluid in motion and a bounding surface when both are at different temperatures [8].

$$q" = -h(T_s - T_\infty) \quad (3)$$

Where;

h is the convective heat transfer coefficient $(W/m^2.^{\circ}K)$

T_s is the surface temperature (°K)

T_∞ is the fluid temperature (°K)

3) Radiation: is the energy emitted by a matter that is at non zero temperature. The energy emmitted by radiation field is transported by electromagnetic waves [6].

$$q" = J - G \quad (4)$$

Where;

J is the radiosity (W/m^2)

G is the irradiation (W/m^2)

$$J = E + \rho G \quad (5)$$

E is the emissive power (W/m^2)

ρ is the reflectivity

4) Energy Balance:

The energy balance is given by the equation below:

$$\frac{\partial}{\partial t}(\rho C_p T) = \frac{\partial}{\partial x}(k\frac{\partial T}{\partial x}) + \frac{\partial}{\partial y}(k\frac{\partial T}{\partial y}) + \frac{\partial}{\partial z}(k\frac{\partial T}{\partial z}) \quad (6)$$

C. *Fluent Modeling*

This case is modeled using Ansys fluent simulation tool. The solution is updated every 0.5 second with 40 iterations per time step for a total flow of approximately 15 mins.

In this study, a heat sink is placed in a fluid domain with one inlet and one outlet. The heat sink will be studied under three different heat fluxes 1250, 2500, and 5000 (W/m^2) at 26 °C as an ambient temperature.

Because our study is a free convection study, pressure inlet is specified as boundary condition for inlet with a pressure gauge equal to zero. Due to symmetry, half of the domain can be studied as shown in Figure 3.

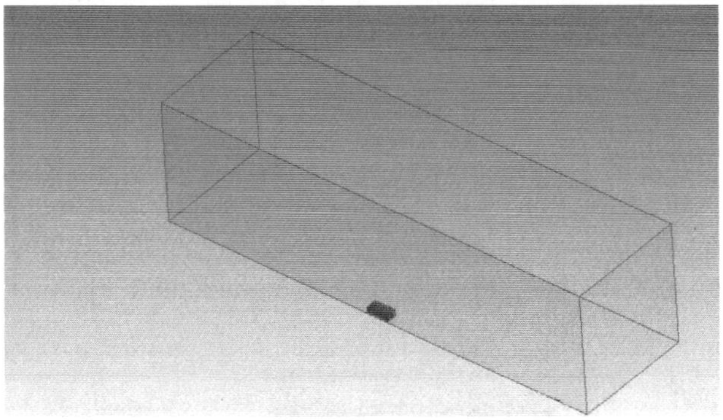

Fig.3. The heat sink and the domain

The domain is divided into 2750000 hexahedral elements as shown in Fig. 4.

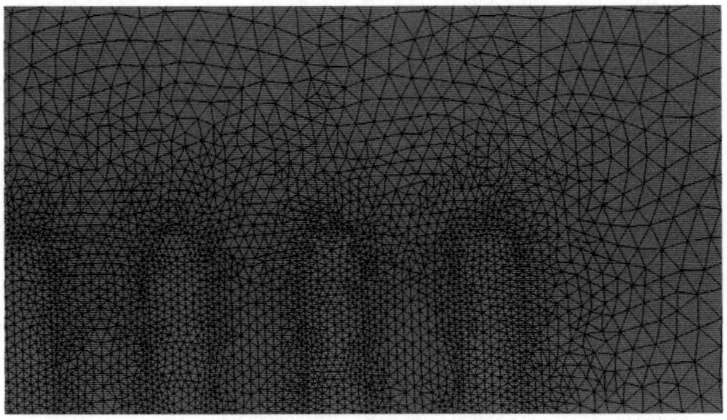

Fig.4. Part of the mesh

IV. RESULTS

As shown in figure 5, the temperature starts to increase from the ambient temperature (300 °K) to the steady state temperature (326.5 °K). At the beginning, the temperature increases significantly, then its rate decreases with time, that's because, when the surface temperature of heat sink increases, the temperature difference with ambient ($T_s - T_\infty$) increases so that the heat rejected by convection and radiation increases with time until reaching steady state and becomes closer to heat generated by the chip (1250 W/m^2). Eventually, heat generated is equal to heat rejected.

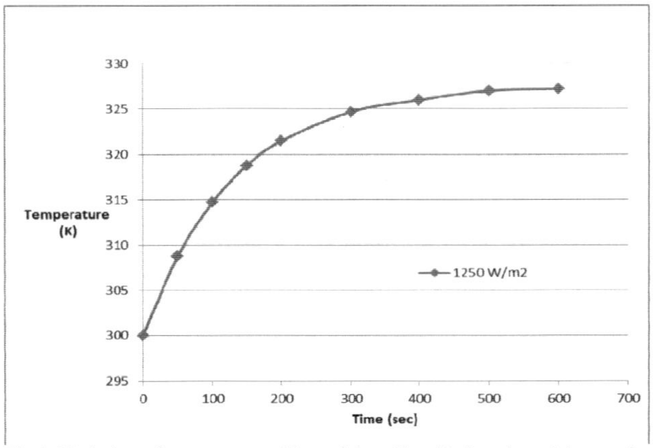

Fig.5. Variation of temperature of heat sink walls with function of time under 1250 (W/m^2).

The time needed for the heat sink to reach steady state temperature is 9.5 min.

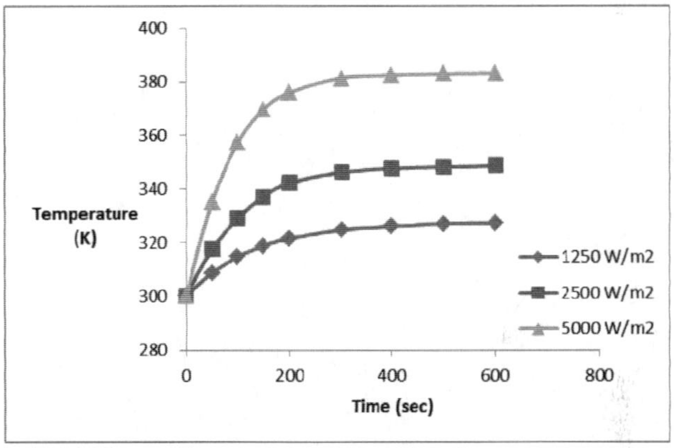

Fig.6. Variation of temperature of heat sink walls with function of time under 1250. 2500 and 5000 (W/m^2).

As shown in figure 6, it is obvious that when the heat flux increases, the steady state temperature increases, until it reaches 348 and 383 °K for 2500 and 5000 W/m^2, respectively.

Note that the temperature reached with 5000 W/m^2 is close to the harmful maximum allowable temperature (393 °K), so the electronic ship fails to work properly, that's why an additional cooling technique is required.

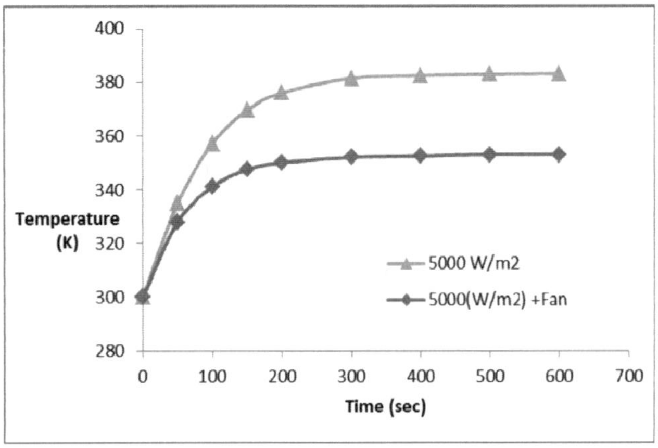

Fig.7. Variation of temperature of heat sink walls with function of time under 5000 (W/m^2) and forced ventilated.

Figure 7 shows clearly that the forced ventilation decreases the temperature of the heat sink to 353 °K instead of 383 °K which is considered an acceptable temperature for a standard operation of electronic chips.

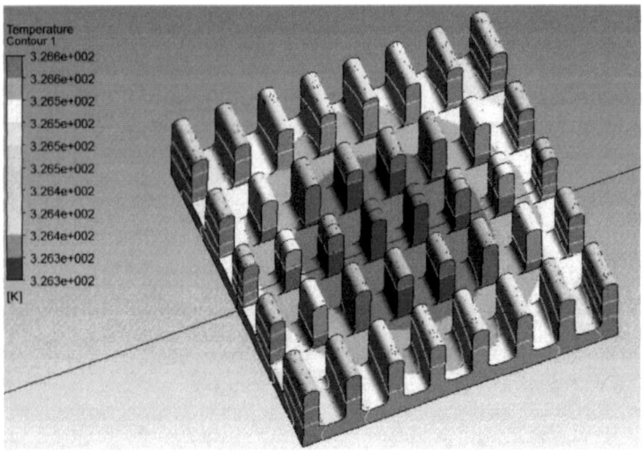

Fig.8. Temperature distribution through the heat sink

Figure 8 clearly shows that the temperature reaches its maximum value at the center of the heat sink, and it decreases gradually when moving away from the center and up to the top of the fins.

From the simulation results shown above, one can conclude that the heat sink might be used as good cooling technique of most of electronic devices. However, traditional heat sinks have some limitations. Based on the later observation, the authors of this paper are currently testing a complementary cooling technique which consists of adding Phase Change Materials (PCMs). The PCM is to be integrated within the heat sink [7].

V. CONCLUSION

This paper aims to simulate the heat distribution over a heat sink using Ansys fluent simulation tool to serve the goal of investigating and studying the effect of integrating phase change materials later on into the heat sink in order to improve the cooling performance and consequently increase the efficiency of electronic chips.

VI. REFERENCES

[1] Hill, M. (n.d.). *Cooling of Electronic Equipments"chapter 15".* Higher education.

[2] Coskun, A. K. 2010. "Energy-efficient variable-flow liquid coolingg in 3D stacjed architectures." *DATE,pp* 111-116.

[3] H. Xie, A. Ali. 2011. "The use of heat pipes in personal computers." *ITHERM* 331-340.

[4] Lasance, Clemens J.M. 2005. "Advanced in high-performance cooling for electronics." *Electric cooling magazine* 2-8.

[5] Lee, S. (n.d.). How To Select a Heat Sink. *Advanced Thermal Engineering.*

[6] Kandasamy, Ravi. 2007. "Transient cooling of electronics using phase change material (PCM)-based heat sinks." *Elsevier*, December 23: 11.

[7] Ahmad hasan, Hassan Hijase, shaimaa Abdelbaqi, Ali Assi, Mohammad Hamdan. "Comparative Effectiveness of Different Phase Change Materials to Improve Cooling Performance of Heat sinks for Electronic Devices." *Applied Sciences*, 2016: 2-3.

[8] Frank P. Incropera David P. Dewitt , Theodore L. Bergman , Adrienne s. Lavine Fundemantals of Heat and Mass Transfer [Book]. - [s.l.] : John willey and sons, 2006.

 https://doi.org/10.1051/matecconf/201817102004

Investigation of the impact of Photovoltaic Sizing and Siting Modeling on Micro-grids Energy Management Optimization

Marwa Abuelnasr
Electrical Power and Machines Department
Faculty of Engineering, Ain Shams University
Cairo, Egypt
marwa.mahmoud@eng.asu.edu.eg

Walid El-Khattam
Electrical Power and Machines Department
Faculty of Engineering, Ain Shams University
Cairo, Egypt
walid.elkhattam@eng.asu.edu.eg

Ibrahim Helal
Electrical Power and Machines Department
Faculty of Engineering, Ain Shams University
Cairo, Egypt
dr.i.helal@gmail.com

Abstract— **The modeling of micro-grid's PV units' and loads' size and distribution along its network could severely affect the accuracy of power losses calculation and; thus the amount of energy to be imported from the main-grid to balance the load. Also, it will affect the micro-grid's optimal energy management results. Therefore, a comprehensive analysis is carried out to assess the impact of the PV's and the load's sizing and sitting (either lumped or distributed). Four case studies are carried out to illustrate the impact of the modeling on the micro-grid's losses and the imported energy from the main-grid and their costs. The obtained results are used to implement the proposed energy management two-single objective optimization functions applying Genetic Algorithm. Then, a fifth case study is carried out to optimize the micro-grid energy management process through an optimal chagrining - discharging scheduling of a storage module. The obtained results, recommendations, and evaluations for choosing a proper sizing and siting modeling and the chagrining - discharging scheduling of a storage module under seasonal variations are reported and discussed.**

Keywords— Energy management optimization, Genetic Algorithm optimization, Micro-grid.

I. INTRODUCTION

The integration of Distributed Energy Recourses (DER) such as wind turbines, Photovoltaic (PV), storage systems, and electric vehicles have increased the complexity of the distribution system's operation. To overcome this complexity, the electric energy consumption, the generation resources and energy storage operation should be optimally managed. This should be carried out in a way that insures distribution system's energy saving, improves energy efficiency, enhances system reliability, reduces greenhouse gas emissions, decreases fuel use and increase the use of Renewable Energy Sources (RES) while minimizing operating cost [1]. This energy management process is carried out through monitoring, controlling, and conserving energy in the whole distribution system.

Existing energy management systems can be divided into centralized and decentralized according to their philosophy of operation. The centralized energy management systems focus on controlling all energy activities in the whole system to reach an optimal energy management operation. However, their main disadvantage is reduced flexibility, as modifications are needed for each additional component (Generators, Storage systems…etc) to be installed in the system and the extensive computational requirements [2]. Due to this lack of flexibility, large numbers of DER in a system are difficult to be handled by the existing centralized control systems. It is difficult or nearly impossible to control the entire system by a single central controller. Thus, to avoid the centralized disadvantages research is being directed at micro-grids (decentralized energy control and management) because of their ability to enhance the reliability of a power system and reduce its environmental impact [3]. To divide existing bulky grids into micro-grids, electrical boundaries for sub-micro-grids should hence be clearly defined: each micro-grid inside the defined boundaries should be controlled as a single entity that match its generation and load so that it can connect to or disconnect from the grid keeping the total load supplied [3].

Various research have been introduced on microgrid energy management based on decentralized energy management system. In [4], optimal allocation of DGs in the PV-based micro-grid was investigated after forecating the PV output power. The Modified Artificial Fish School algorithm was implemented on a cost function to optimaly manage energy in a PV based micro-grid. In [5], a nonlinear energy management system was applied to an islanded microgrid with storage system. Genetics Algorithm-based (GA) hybrid optimization algorithm was implemented on two levels in [6]. First, it

This work was supported in part by the short term fellowship of the Science & Technology Development Fund (STDF), Ministry of Higher Education and Scientific Research, Egypt.. [Project No.6463: Efficient Protection Systems for Microgrids.]

applied nonlinear programming for modeling the energy management problem. Second, a hybrid solution was implemented to solve the microgrid energy management problem under study. A micro-grid with renewable and non-renewable DGs was tested in [7] through a proposed distributed energy management strategy to reach the optimum energy management operation taking into consideration the distribution network constraints. In [8], the control of the power flow in the micro-grid was controlled in order to minimize power losses, energy cost, and dissatisfactions of the customers. It proposed a distributed energy management of micro-grids where both supply-side and demand side management were considered. In [9], a decentralized partially-observable Markov decision process was proposed to minimize operational costs of the micro-grid through coordinating the power production of DG sources and energy storage system. An energy management algorithm was presented in [10] in order to optimize the energy storage applications within micro-grids. While [11], GA was implemented to minimize energy cost and emissions of a micro-grid with different DGs and energy storage systems in both winter and summer seasons. However, it did not go into the details of each of the micro-grid's decision variables.

In this paper, the energy management problem of a PV–energy storage-based grid connected micro-grid is investigated through a proposed optimization process to minimize energy loss and cost. The analysis take into consideration; investigating the impact of sizing and sitting modeling of PV and load on the accuracy of the calculations, studying energy storage charging-discharging schedule, incorporating the impact of seasonal variations and customers behavioral changes on the demand.

II. PROBLEM FORMULATION

In this section, the problem under study's mathematical formulation; two objective functions and their associated operational constraints, is introduced. In addition, a detailed description of the system under study and the decision-making tool are presented.

A. Mathematical Problem Formulation

This section discusses the detailed mathematical formulation of energy management optimization problem under study.

1. Objective Functions:

Objective Function 1:

The first objective is to minimize the micro-grid energy loss that is calculated during the specified study period using its total number of hours, as follows [1]:

$$Min. \quad E_{loss} = 0.5x \sum_{h=1}^{h_{tot}} \sum_{i=1}^{n} \sum_{j=1}^{n} \left(G_{ij}^{h} x \left[\left(V_i^h\right)^2 + \left(V_j^h\right)^2 - 2xV_i^h x \right. \right. \tag{1}$$

Where;

V_i is the per unit voltage at bus i,

G_{ij} is the conductance of the line connected between buses i and j,

δ_i & δ_j are the power angles of buses i & j,

n is the total number of buses,

h is the hour counter,

h_{tot} is the total number of hours during the study period.

Objective Function 2:

The second objective is to minimize the total amount of energy received by the micro-grid from the main grid represented by its total energy cost. The total energy cost is given by the following:

$$Min. \quad Cost = a * \sum_{h=1}^{h_{tot}} c_{grid}^{h} x \, E_{grid}^{h} \tag{2}$$

Where;

a is an adjustment factor

c_{grid} is the grid energy cost/hour

E_{grid} is the total amount of energy supplied by the grid

2. Constraints

The operation of the micro-grid under study is subject to the following constraints [1]:

- *Power Balance Equations:*

$$P_{i_\phi}^{h} = \sum_{j=1}^{n} \left|V_{i_\phi}^{h}\right| \left|V_{j_\phi}^{h}\right| \left(G_{ij_\phi}^{h} \cos\left(\theta_{ij_\phi}^{h}\right) + B_{ij_\phi}^{h} \sin\left(\theta_{ij_\phi}^{h}\right)\right) \tag{3}$$

$$Q_{i_\phi}^{h} = \sum_{j=1}^{n} \left|V_{i_\phi}^{h}\right| \left|V_{j_\phi}^{h}\right| \left(G_{ij_\phi}^{h} \sin\left(\theta_{ij_\phi}^{h}\right) - B_{ij_\phi}^{h} \cos\left(\theta_{ij_\phi}^{h}\right)\right) \tag{4}$$

$\forall i \in \{1,2, \dots n\}, h \in \{1,2, \dots h_{tot}\} \;\&\; \emptyset \in \{a, b, c\}$

Where;

P and *Q* are the injected kW and kVAr respectively,

B_{ij} is the susceptance in the bus admittance matrix that corresponds to the i[th.] row and j[th.] column,

θ_{ij} is the difference in voltage angles between the ith and j[th.] buses, and

$\emptyset$ is the phase number.

- *Voltage Magnitude Limits* [1]:

$$V_{min} \leq \left|V_{i_a}^{h}\right|, \left|V_{i_b}^{h}\right|, \left|V_{i_c}^{h}\right| \leq V_{max}, \forall i \in \{1,2, \dots n\} \tag{5}$$

- *Lines Thermal Limits* [1]:

$$\left|S_j^h\right| < \left|S_{j\ max}^{h}\right| \qquad \forall i \in \{1,2, \dots N_L\} \tag{6}$$

Where;

S_j is the line capacity

N_L is the total number of micro-grid's lines.

- *Storage model control tool:*

The storage charging/discharging schedule is controlled by a controller that follows a load shape. The energy output of the storage model is expressed by:

$$E_s = E_{stored} + \sum_{h=1}^{h_{tot}} c_s^h E_{so} \tag{7}$$

Where;

E_{stored} is the initial stored kWh in the battery,

E_{so} is the storage rated kWh,

E_s is the total stored kWh, and

c_{si}^{h} is the load curve value for the storage at instant h.

B. *System Description*

The micro-grid test system is shown in Figure 1. It is part of the IEEE 123-bus three-phase unbalanced test system [12] according to an adequacy criteria explained in [3]. The micro-grid is assumed to have scattered 1000 kW and 858 kVAr load and five-solar based microFITs [13] rated at 700 kW and a storage unit rated at 250 kW, 1000 kWh. The tariff of the imported energy from the main-grid is calculated based on [14]. The micro-grid is assumed to be equipped with storage module with schedule control tool. The energy management system makes hourly generation plan for the next 24 hours to minimize the total operating cost based on the forecast power production of PVs and loads [2]. All PVs are assumed to work at unity power factor as the focus of this study is on the active power scheduling.

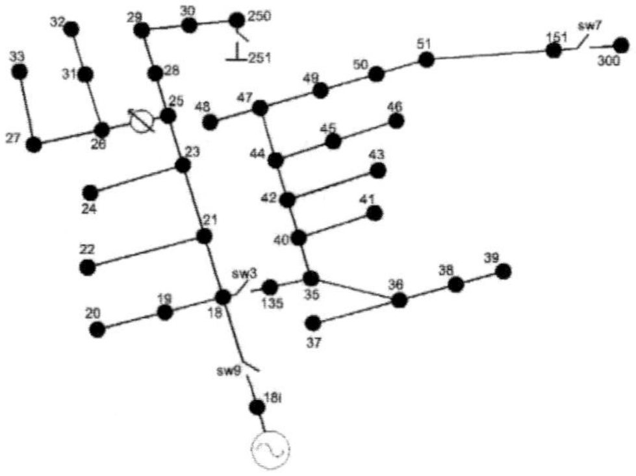

Figure 1 The micro-grid under study

C. *Decision-Making Tool*

The system under study is tested for single objective optimization technique each at a time. Genetic Algorithm (GA) was implemented to generate the required sets of solutions. GA relative to other Artificial Intelligence algorithm is less complex, more straightforward, more flexible, and offers a close-to-global optimum solution in a much shorter time. In the case of single objective, each chromosome is a proposed solution set (e.g. a charging or discharging of the storage model). If all constraints are satisfied, the system objective function is calculated and becomes the fitness value. These steps are repeated until a stopping criterion is reached. For this research, the stopping criterion is the generation number, which is provided by the operator.

III. APPLICATIONS

In this paper, four case studies are carried out to investigate the impact of modeling both load and PV (i.e. lumped and/or scattered) sitting and sizing on the value of the micro-grid's losses and energy received from the main grid as follows:

- *Case study (1):* A lumped load with a varying capacity lumped PV at the same bus
- *Case study (2):* A lumped load with a varying capacity lumped PV at two different buses
- *Case study (3):* Scattered loads with a varying capacity lumped PV
- *Case study (4):* The micro-grid with scattered loads and scattered 5 PVs

These analyses are carried out applying load flow analysis using OpenDSS software [15] representing a whole day (24 hours) of a typical day in the winter and a typical day in the summer. A stochastic data model, developed in [2], was utilized to forecast the load values and microFITs power output during the study periods.

After this investigation, for each season, the system is analyzed for two single-objective optimization problems namely minimizing the micro-grid's cost of; losses and energy received from the main grid. Case study (5) is carried out to optimization the micro-grid that poses scattered loads and scattered 5 PVs and an added 1 storage models. The storage module charging and discharging process is optimized to achieve the minimum loss and energy received from the main grid.

Case study (1): A lumped load with a varying capacity lumped PV at the same bus

The system is simulated with a single lumped load (1000 kW and 858 kVAr) and a single lumped PV unit with varying capacity at the same bus, Bus 30. For both summer and winter simulations results as in Tables 1.a and b respectively, the micro-grid's losses are decreased until a certain capacity of PV and then starts to increase again when the PV kWh is larger than that of the load. In the meanwhile, as the PV capacity increase, the amount of energy imported from the main grid decreases. Thus, the imported main grid energy's cost decreases. Also, it is shown that the PV capacity required in winter in more than that for summer due to the seasonal sun irradiance differences. Table I.b shows a negative cost that represents an export of energy from the micro-grid to the main-grid due to excessive PV generation than the micro-grid load. By comparing the losses between seasons, it is clear that the summer losses is higher than that of the winter as the summer loading is higher than the winter one as shown in Figures 2 and 3.

Table I.a Micro-grid losses and imported main grid energy cost for Case study (1) in winter

PV (kW)	Winter	
	Cost ($/kWh)	Losses (kWh)
0	1714.1	81
500	1674.7	77
1000	1582.7	73
2000	1451.4	67
3000	1320.4	64
4000	1189.5	63
5000	1058.8	64
6000	928.4	68
7000	798.1	73

 https://doi.org/10.1051/matecconf/201817102004

Table I.b Micro-grid losses and imported main grid energy cost for Case study (1) in summer

PV (kW)	Summer	
	Cost ($/kWh)	Losses (kWh)
0	2088.8	123
100	2003.5	117
200	1918.2	112
300	1833.0	107
400	1747.8	103
500	1662.7	99
1000	1237.7	85
2000	391.7	88
3000	-449.7	129

Figures 2 and 3 show the change of the PV capacity with respect to the load curve of the lumped load in both winter and summer seasons respectively.

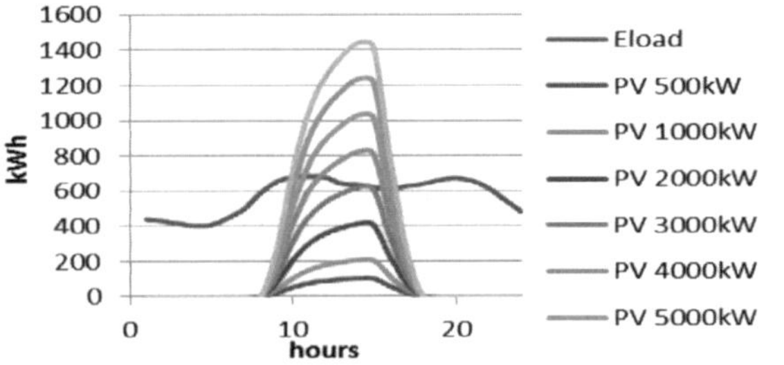

Figure 2 Case study (1): Micro-grid's PV and load energies in winter

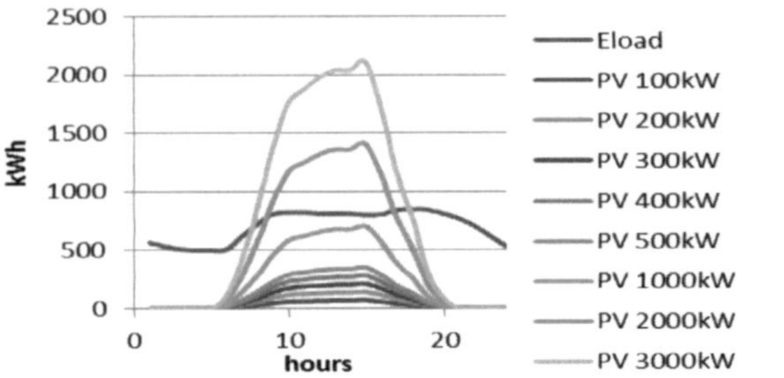

Figure 3 Case study (1): Micro-grid's PV and load energies in summer

Case study (2): A lumped load with a varying capacity lumped PV at the different buses

Like in Case study (1), the simulation is carried out with PV installed at a different bus randomly, Bus 40, other than that of the load while keeping the same conditions in Case study (1) unchanged. It is realized that by increasing the PV capacity the losses kept constant until at a certain capacity of PV and then it start to increase. In the meanwhile, as PV capacity increases, the imported main-grid energy cost keeps dropping as shown in Tables II.a and b.

By comparing the results obtain in Tables 1 and 2, it is clear that having PV on different bus than the load will increase the load flow in the micro-grid and thus increases the losses and the amount of imported energy from the main-grid.

Table II.a Micro-grid losses and imported main grid energy cost for Case study (2) in winter

PV(kW)	Winter	
	Cost ($/kWh)	Losses (kWh)
0	1714.1	81
400	1661.7	81
500	1648.6	81
1000	1583.1	81
2000	1452.2	82
3000	1321.3	85

Table II.b Micro-grid losses and imported main grid energy cost for Case study (2) in summer

(PV kW)	Summer	
	Cost ($/kWh)	Losses (kWh)
0	2088.8	123
100	2003.8	123
200	1918.9	123
300	1834.0	123
400	1749.0	124
500	1664.1	125

Figures 4 and 5 show the change of the PV capacity with respect to the load curve of the lumped load in both winter and summer seasons respectively.

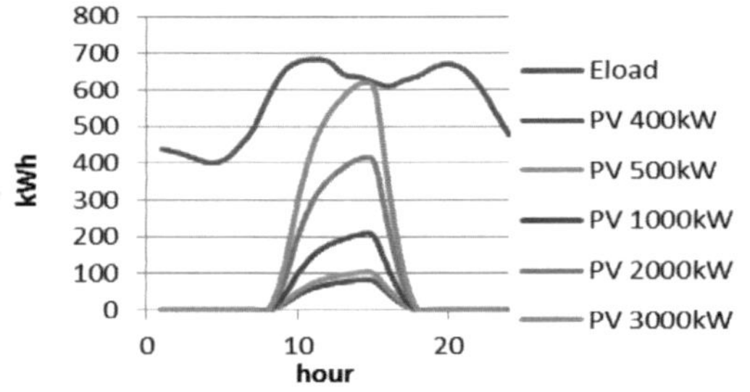

Figure 4 Case study (2): Micro-grid's PV and load energies in winter

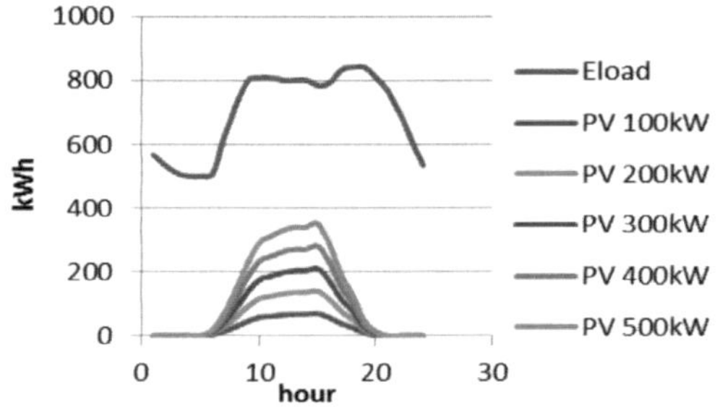

Figure 5 Case study (2): Micro-grid's PV and load energies in summer

Case study (3): Scattered loads with a varying capacity lumped PV

This simulation tests five scattered loads with same kW and kVAr as in Case study (1) installed on Buses-19, 28, 37, 43, and 50 while the PV is installed at Bus-40. Results shown in Tables III.a and b indicate that the losses decrease then increase again, while the imported energy's cost is reduced as long as the PV capacity increases. Figures 6 and 7 show the change of the PV capacity with respect to the load curve of the total load in both winter and summer seasons respectively.

Table III.a Micro-grid losses and imported main grid energy cost for Case study (3) in winter

PV kW	Winter	
	Cost ($/kWh)	Losses (kWh)
100	1705.7	63
500	1653.2	63
1000	1587.6	62
2000	1456.5	62
3000	1325.5	62
3500	1260.0	62
4000	1194.5	63

Table III.b Micro-grid losses and imported main grid energy cost for Case study (3) in summer

PV kW	Summer	
	Cost ($/kWh)	Losses (kWh)
100	2010.4	95
500	1670.2	93
1000	1245.3	93
1100	1160.4	93
1200	1075.5	94
1300	990.6	94
1400	905.8	95
1500	820.9	96
2000	397.0	104

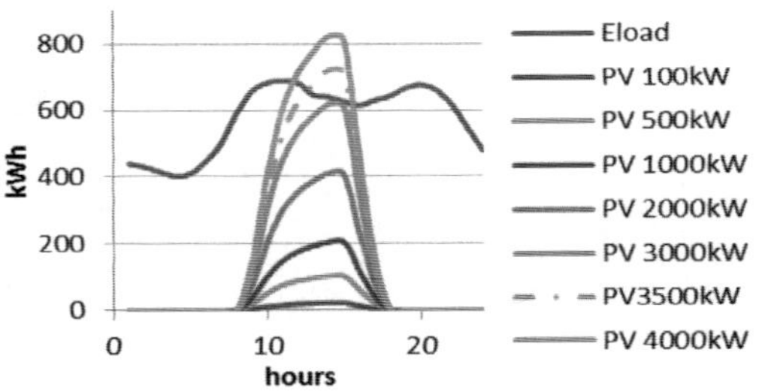

Figure 6 Case study (3): Micro-grid's PV and load energies in winter

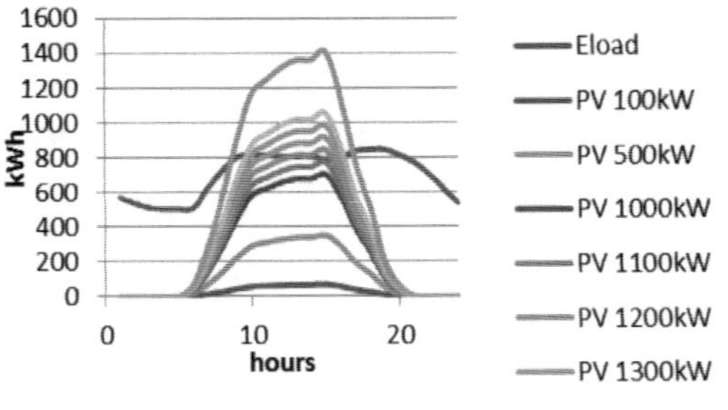

Figure 7 Case study (3): Micro-grid's PV and load energies in summer

Case study (4 (Benchmark)): Scattered loads with a varying capacity scattered PV

The results obtain from this simulation showed almost similar results as in Case study (3).

From the above shown results obtain in Case studies (1) to (4), it can be concluded that near outputs can be achieved if the modeling of micro-grid loads and PV are scattered or lumped at the same bus. However, Case study (3) shows more accurate results than those of Case study (1). Therefore, the rest of the analysis in this paper will be carried on the concept of scattered loads and PV with a capacity that obtained in the winter season in Case study (3), i.e. 3500 kW.

Case study (5): Optimization of the micro-grid with scattered 5 PVs and 1 storage module

The proposed two-single objective function optimization algorithm using GA technique is implemented to minimize the micro-grid losses and imported energy from the main-grid. This minimization is carried out through optimizing the storage module charging and discharging schedule. Tables IV.a and b show the optimal values in both seasons.

Table IV.a Optimized micro-grid losses and imported main grid energy cost for Case study (5) in winter

PV kW	Winter	
	Cost ($/kWh)	Losses (kWh)
0	1900.0	29
3500	976.4	25

Table IV.b Optimized micro-grid losses and imported main grid energy cost for Case study (5) in summer

PV kW	Winter	
	Cost ($/kWh)	Losses (kWh)
0	2319.5	44
3500	-1114.3	38

In winter season, the optimization was able to reduce the losses and cost of imported energy by 60% and 23% respectively when compared with results obtained in Table IV.a. Similarly, in summer, the optimization was able to reduce the losses by 41% while and provides a saving by exporting energy to the main-grid by 200% when compared with results obtained in Table IV.b. Figures 8 and 9 show the obtained optimal results.

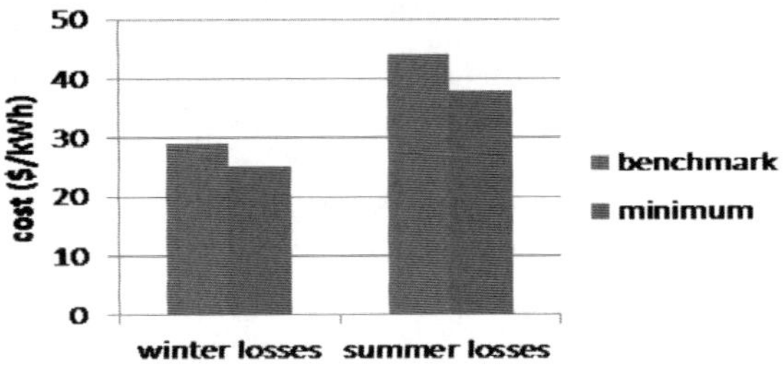

Figure 8 Optimization of the micro-grid's

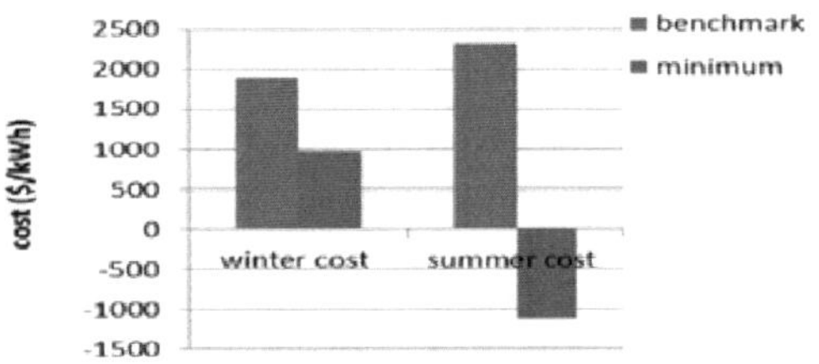

Figure 9 Optimization of the cost of the imported energy in winter and summer

IV. CONCLUSIONS

The impact of sizing and sitting (scattered or distributed) modeling of a micro-grid's PV and loads was evaluated. Results show that under the current micro-grid's structure; the scattered modeling of PV and loads along the micro-grid's network provided near results of having a micro-grid's lumped PV and load installed at the same bus. However, modeling PV and load as lumped values at different busses reported far results. Therefore, accurate modeling will impact the calculated values of power losses as well as the amount of energy to be imported from the main-grid. This obtained results of PV and load sitting as well as the PV sizing that covers the micro-grid load under both winter and summer variations were used to optimally solve the micro-grid's energy management problem. A proposed two-single objective optimization functions via applying a GA is implemented to minimize the micro-grid losses and cost of imported energy from the main grid through optimal scheduling of charging and discharging of a storage module. Results show that the optimization was able to save a huge amount of micro-grid losses in both winter and summer. In the meanwhile, the optimization was able to reduce the amount of imported energy from the main-grid in winter, while was capable to export energy to the main-grid in summer through optimal charging – discharging scheduling of the storage module.

REFRENCES

[1] H. A. Mostafa, R. E. Shatshat and M. M. A. Salama, "Optimal Distribution Systems Operation Using Smart Matching Scheme (SMS) for Smart Grid Applications," *IEEE Transactions on Smart Grid,* vol. 5, no. 4, pp. 1938-1948, 2014.

[2] H. A. Mostafa, "Zonal Energy Management and Optimization System (ZEMOS) for Smart Grid Applications," Ontario, Canada, 2014.

[3] M. E. Nassar and M. M. A. Salama, "Adaptive Self-Adequate Microgrids Using Dynamic Boundaries," *IEEE Transactions on Smart Grid,* vol. 7, no. 1, pp. 105-113, 2016.

[4] Ke Huang; Xin Wang; Liming Wang, "Optimal energy management of grid-connected photovoltaic micro-grid," IEEE International Conference on Cyber Technology in Automation, Control, and Intelligent Systems (CYBER)*, pp. 234-239, 2015.*

[5] Zhiqiang Guo, Deshang Sha, and Xiaozhong Liao, "A Wireless, Decentralized and Nonlinear Energy Management Strategy Using PCC Frequency for Islanded AC Microgrids," *IEEE Applied Power Electronics Conference and Exposition - APEC 2014*, pp. 3136 – 3142, 2014.

[6] H. Li, C. Zang, P. Zeng, H. Yu and Z. Li, "A genetic algorithm-based hybrid optimization approach for microgrid energy management," *Cyber Technology in Automation, Control, and Intelligent Systems (CYBER)*, 2015.

[7] W. Shi, X. Xie, C.-C. Chu and R. Gadh, "Distributed Optimal Energy Management in Microgrids," *IEEE Transactions on Smart Grid,* vol. 6, no. 3, pp. 1137 - 1146, 2015.

[8] W. Shi, X. Xie, C. C. Chu and R. Gadh, " Distributed Optimal Energy Management in Microgrids," *IEEE Transactions on Smart Grid,* vol. 6, no. 3, pp. 1137 - 1146, 2015.

[9] J. Wu and X. Guan, "Coordinated Multi-Microgrids Optimal Control Algorithm for Smart Distribution Management System," *EEE Transactions on Smart Grid,* vol. 4, no. 4, pp. 2174 - 2181, 2013.

[10] R. Zamora and A. K. Srivastava, "Energy management and control algorithms for integration of energy storage within microgrid," in *Industrial Electronics (ISIE), 2014 IEEE 23rd International Symposium*, 2014.

[11] M. Elsied, A. Oukaour, H. Gualous, R. Hassan and A. Amin, "An advanced energy management of microgrid system based on genetic algorithm," in *Industrial Electronics (ISIE), 2014 IEEE 23rd International Symposium*, 2014.

[12] "IEEE Distribution Planning Working Group Report, Radial Distribution Test Feeders," *IEEE Transactions on Power Systems,* vol. 6, 1991.

[13] [Online]. Available: http://microfit.powerauthority.on.ca/introduction

[14] [Online]. Available: http://www.ieso.ca/power-data/price-overview/hourly-ontario-energy-price

[15] [Online]. Available: https://sourceforge.net/projects/electricdss/

Hidden High Voltage Safety Risks for Parallel High Voltage Transmission Lines

M. Nassereddine, J. Rizk, M. Nagrial, A. Hellany
School of Computing, Engineering & Mathematics
Western Sydney University, Penrith Australia
mnassereddine@ieee.org

N. Moubayed, N. Karami
LaRGES, CRSI, Faculty of Engineering
Lebanese University, Tripoli, Lebanon
nmoubayedi@ieee.org, nabilkarami@hotmail.com

***Abstract*--** **High voltage (HV) infrastructures market is growing due to the corresponding growth in industries and population. To ensure continuous and reliable electrical power supply, existing substation and transmission lines are being upgraded to accommodate the additional load requirements. These upgrades involve up-rating the existing transmission lines or the installation of new lines. To save on easement cost and reduce the environmental impacts, transmission lines are occupied the same easement or path. This parallel option introduces the risk of induced voltage which could reach an unsafe condition and jeopardize the safety of works and people. This paper analysis and highlight the hidden risk associated with two parallel transmission lines that connected the same high voltage substation. The theoretical study which is supported by the case study shows the high risk potential tempering with the OHEW on the isolated circuit while the other one is still energized**

***Index Terms*—Earth Potential Rise, Fault Current, High Voltage Transmission, Induced Voltage.**

I. Introduction

Electrical power forms an essential element for human daily ongoing activities. Numerous happenings would have been impossible without the aid of the electrical power. The generated electrical power, at the power station, passes through numerous stages to reach the consumers. One of these stages is the transmittal of the generated energy using transmission lines.

The transmission lines are essential to transfer the electrical energy between numerous high voltage infrastructures such as power station, bulk supply point, transmission substation and zone substation. To reduce the easement cost and the environmental impact, where possible, these transmission lines occupy the same easement route. Figure 1 shows the transmission lines arrangement for the main energy supplier to NSW, Australia, electrical utilities [1]. It is clearly shown within figure 1 the numerous transmission lines run parallel and connect the same high voltage infrastructures. It is worth mentioning, two transmission lines that connects the same substation are usually physically segregated for reliability, serviceability and availability purposes.

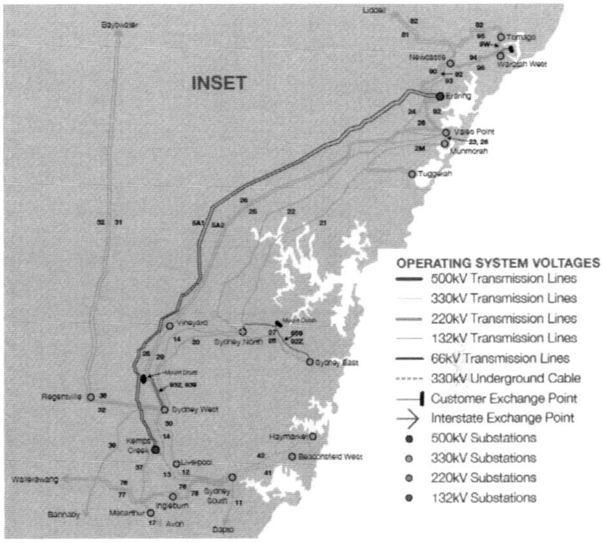

Fig. 1: Transgrid electrical network map for NSW Australia

Induced voltage on parallel conductive object to an energized transmission line could reach an unsafe condition under fault and load conditions [2]. Therefore, parallel transmission lines introduce the risk of induced voltage on each other conductive conductors which include the overhead earth wire (OHEW).

The OHEW form an essential element for the transmission lines when it comes to fault current distributions and lightning protections. The maintenance of high voltage transmission line includes the OHEW. As the OHEW is earthed at each pole, workers assume it is safe to perform the job as long its circuit is isolated and earthed. This theory overlook the hidden risk exists on the isolated and earthed transmission under fault condition on the energized parallel feeder.

The works in this paper analysis and highlight the hidden risk associated with two parallel transmission lines that connected the same high voltage substation. The theoretical study which is supported by the case study shows the high risk potential tempering with the OHEW on the isolated circuit while the other one is still energized.

II. Analysis

The flow of phase conductor current creates a magnetic field which induces a voltage on any conductive object running parallel to the current. The OHEW current is driven by the split factor and the included voltage concept [2-7]. Figure 2 represents a transmission lines with continuous OHEW that connect both substations [8].

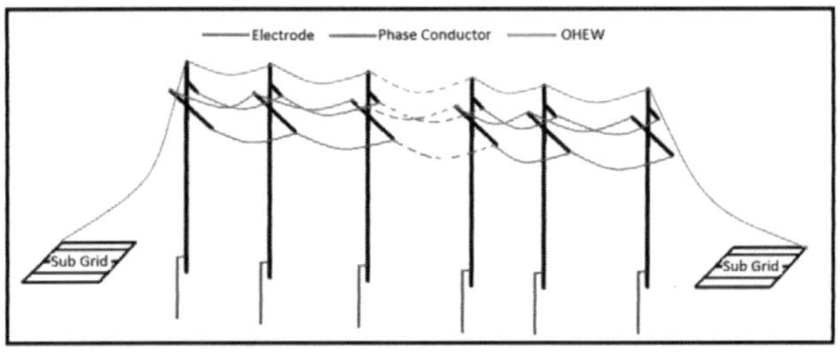

Fig. 2: Transmission lines with continuous OHEW between two substations

Figure 3 shows the electrical circuit for fault current distribution under the continuous OHEW shown in 2 [8].

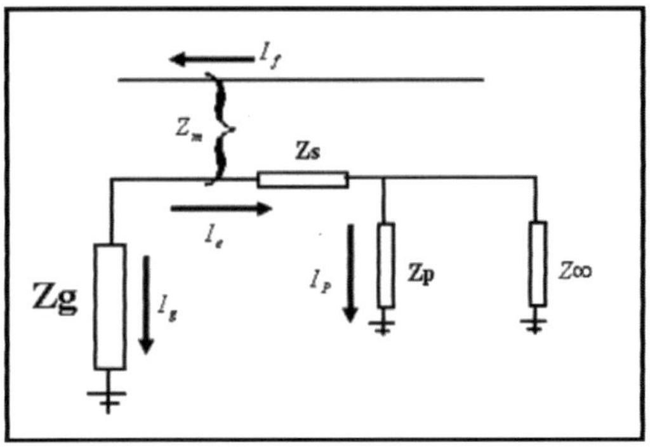

Fig. 3: Split Factor circuit analysis

where,

Z_s is the OHEW self-impedance for the average span,

Z_p is the pole earth grid resistance,

Z_∞ is the OHEW input impedance [8],

Z_m is the mutual impedance between the faulted phase and the OHEW,

I_f is the fault current,

I_g is the earth grid current,

I_P is the transmission pole earth grid resistance,

I_e is the OHEW return current.

Equations 1, 2 and 3 can be used to determine the OHEW section current and the transmission pole earth grid current along the transmission lines [9]

$$I_{Pn} = \left[\frac{Z_P}{Z_\infty + Z_P} \right]^{n-1} I_{P1} \tag{1}$$

$$I_{P1} = \frac{\delta_e Z_\infty - Z_m}{Z_\infty + Z_P} I_f \tag{2}$$

$$I_n = \delta_e I_f - \sum_{1}^{n-1} I_{Pi} \tag{3}$$

where, n is the number of pole and δ_e OHEW split factor which can be computed using equation 4

$$\delta_e = \frac{Z_g + Z_m}{Z_\infty + Z_g} \tag{4}$$

Based on the above equations and previous research, figures 4 and 5 show the current behavior within the OHEW and transmission pole earth grid current under fault condition.

The OHEW section current dives until reach a limit where it maintains its current magnitude within the OHEW. The pole earth grid current dives until reaches the zero value. Based on these two figures, there is no pole earth potential rise (EPR) in the middle of the transmission line under substation fault. Furthermore, the OHEW section current is at its minimum value in the middle of the transmission line.

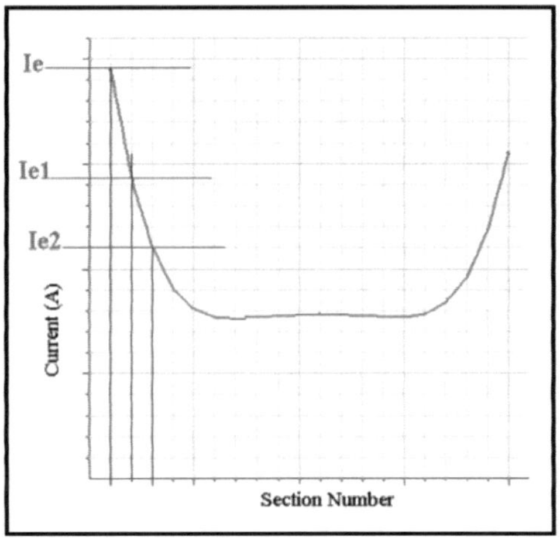

Fig. 4: OHEW section current for a continuous OHEW

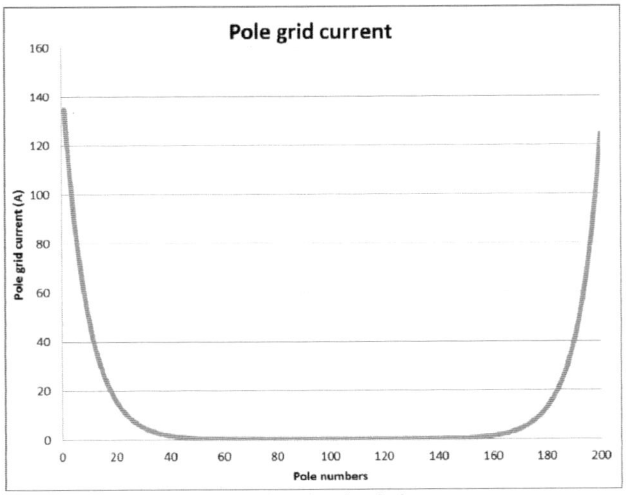

Fig. 5: Pole earth grid current under substation fault

When two transmission lines are running parallel and connect the same substations, the OHEW on the second feeder acts as a second OHEW for the first line due to the followings:

- Both OHEW are bonded at both substations
- Both OHEW are running parallel to the energized transmission line

Figure 6 represents two parallel transmission lines. Both lines has continuous OHEW between substations. When a fault fed by TL1 at Sub1 while TL2 is isolated, the OHEW on TL2 act as a second OHEW for TL1 and aid with the return current.

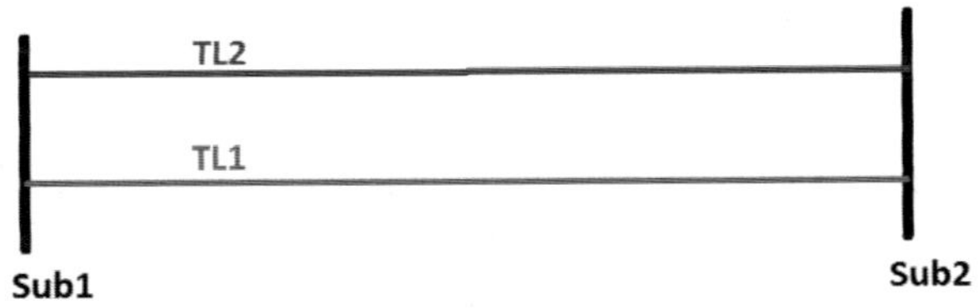

Fig. 6: OHEW for the parallel transmission lines

The return current in OHEW on TL2 behaves as per figures 4 and 5 under continuous OHEW condition. Therefore, based on this information, works at the middle of the transmission line has minimum exposure to EPR under substation fault.

III. Hidden Risk

As previously discussed, the parallel arrangement of the transmission lines introduces the induced voltage risk. Add to this fact that both transmission lines are connected to the same substation which traduced additional return current into the connected OHEW or earthed phase conductor.

Working on high voltage the transmission line involves the following steps from the electrical operator:

- Isolate the transmission lines at all viable sources
- Earth the phase conductors of the transmission lines.

The earthing will depend on the location of the work site. If the work site is located near the substation, an earth will be placed at the substation switchgear as well as the fields. If the work site located in a remote area of the substation, a field earth will be installed at both side of the work location. It is worth mentioning that the working earth will be installed on the phase conductors only and not on the OHEW as it is already earthed at each pole. Figure 7 shows an example of working earth location along a transmission line.

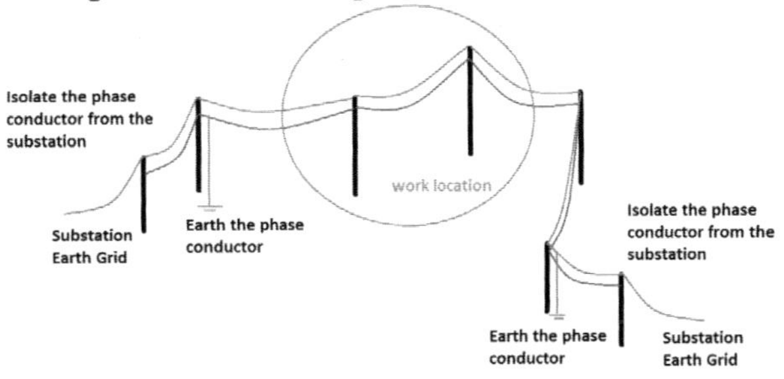

Fig. 7: Field earth location for a working site along the transmission lines

If a substation fault fed by the parallel transmission line, the phase conductor on the isolated line is exposed to an induced voltage while the OHEW of the isolated line is exposed to the split factor concept.

If the proposed scope of work involves breaking the existing OHEW on the isolated line, the behavior of the OHEW section current changes to the one discussed in figure 4 for both layout, finite and infinite conditions. Based on the published work [8], for a non-continuous/broken OHEW the current behave as per figure 8 [8]. The highest current will be discharged on the last few poles; therefore, the worst case scenario will be represented on the last few poles.

- Infinite System: When the OHEW input impedance at the fault location cannot capture the entire length of the system and therefore only a section of the OHEW system is considered for Z_∞
- Finite System: When the OHEW input impedance at the fault location can capture the entire length of the system.

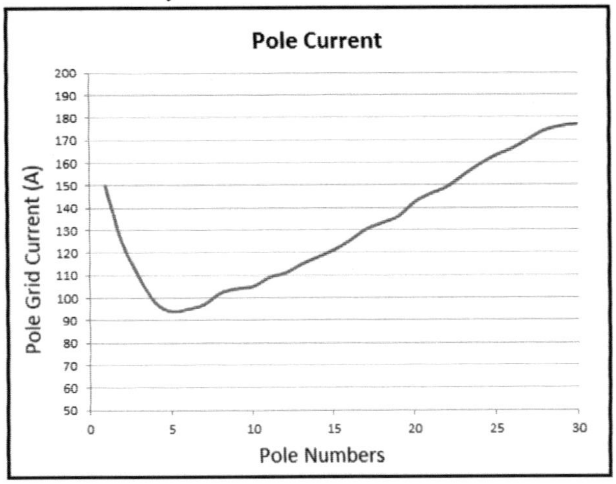

Fig. 8: Pole current behavior for non-continuous/broken OHEW infinite condition [8].

Grounded by the above information, workers touching the broken OHEW (to do maintenance for example) will be exposed to the highest EPR along the line. This hidden EPR on the OHEW, that usually considered the safest assets as it is earth at each pole, has the potential to cause serious damage to workers.

IV. ANALYTICAL DISCUSSION

Based on the information provided in this section, the following can be concluded:

- Under continuous OHEW condition, if the worksite is located in the middle of the transmission lines, the pole EPR under substation fault is at its minimum value
- If the scope of works requires breaking the existing OHEW, the higher pole EPR is located at the broken OHEW location.
- This change in conditions, introduce a high safety risk situation that required mitigations

The mitigation process can take one obtained using one of the followings:

- Isolate both transmission lines, usually this not an option due to network reliability and availability.
- Disconnect the OHEW at the substation to eliminate the split factor concept and rely only on the induced voltage risk.
- Prior breaking the OHEW, create a short circuit that bypass the work location. This could be easily established by bonding the OHEW to the phase conductor both side of the work location as shown in figure 9

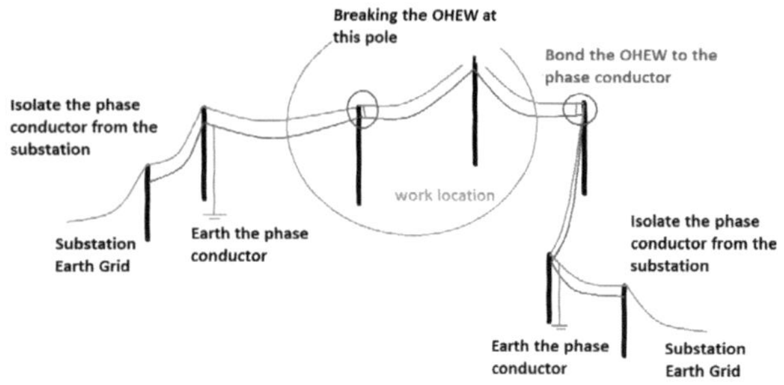

Fig. 9: Bonding the OHEW to the phase conductor

The established bond ensure the returned OHEW current has a continuous pass back to the source and ensure the lowest pole EPR is maintained at this location.

It is worth mentioning, this option is the proposed paper preferred option to mitigate the high risk hidden risk for parallel transmission lines.

V. CASE STUDY

The case study is based on two 132kV feeders running parallel for 20km and connects two zone substations. The followings form the design inputs for both feeders.

- Sub A earth grid resistance is 0.1 ohm
- Sub B earth grid resistance is 0.1 ohm
- Average span on both feeders (FDRA, FDRB) is 10m
- Transmission poles earth grid resistance is 10 ohms
- Single line to ground fault at substation is 10kA
- Delta configuration on the transmission pole
- Separation between the two feeders is 10m
- Separation between the phase conductor and OHEW is 2.4m
- 19/3.25 is the OHEW on both feeders
- Primary clearance time is 0.3S
- Soil resistivity of the area is 50ohm.m

The first assessment is completed by considering the presence of only one feeder. FDR A transmission pole earth grid current for a substation fault supplied by FDR A is computed using equation 1 which illustrated in figure 10. Equation 3 is used to compute the OHEW shunt current. Figure 11 shows the computed result.

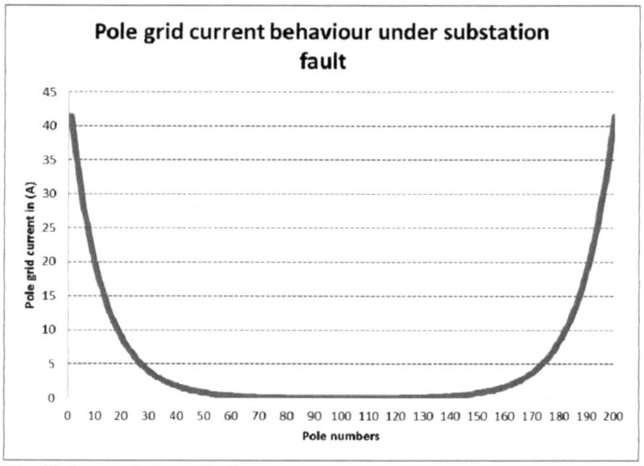

Fig. 10: Transmission pole shunt current

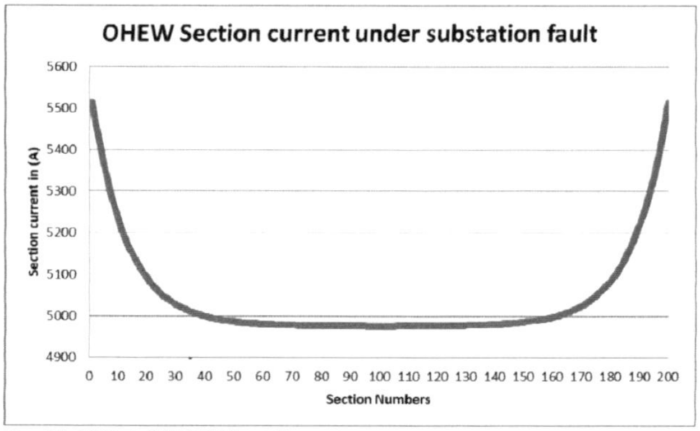

Fig. 11: FDR A OHEW Shunt Current

CDEGS Engineering software is used to compute the transmission pole shunt current on both feeders OHEW when one feeder is under fault at the substation. The followings represent the results:

Total substation fault current supplied by FDR A is 10000 A
Total OHEW on FDR A is 5148.3 A angle: 2.719 degrees
Total OHEW on FDR B is 1263 A angle:1.591 degrees

The simulation shows that the current magnitude in FDR B OHEW is high. The mutual impedance between the faulted phase and the FDR B OHEW is computed to be [8]:

$$Z_{Mutual-\mathrm{Im}pdeance} = 0.0494 + j0.2496\Omega .km$$

The magnitude of the mutual impedance that governs the remaining current into the OHEW (this current will return back to the source using the OHEW) is computed to be 0.329. The equations for the calculation can be found in the author previous published works [8, 9]:

The simulation, which is presented in figure 12, shows that 615A will remain in the OHEW. During a break in the OHEW, this 615A will be discharged to the ground and cause the EPR at the break location. This value along with the resistance will create a high potential that could jeopardize the human safety.

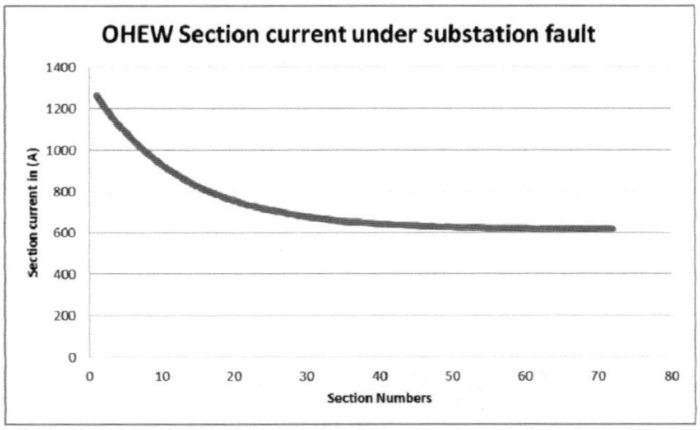

Fig. 12: OHEW section current for FDR B when FDR A is under substation fault

VI. Case Study Mitigation

The consequences of the hidden risk, as detailed in the case study, are high. The high potential could lead to human fatality or permanent damage. This risk is unacceptable and mitigations should be implemented. The paper proses the following mitigation to ensure the risk is reduced:

- When working requires breaking the continuous OHEW, a parallel connection should be established both end of the break.
- The parallel can be achieved as shown in figure 9.

This mitigation ensures the voltage at the work location, under substation fault, will be neglected as the current will be maintained within the OHEW and its parallel section. It is assumed that the mutual coupling between the faulted phase and the small parallel section of the OHEW remain the same as the OHEW.

VII. Conclusions:

The paper addresses the fault current distribution within the OHEW under fault condition. It also shows the fault current behavior under continuous OHEW for a two feeders running parallel within the same easement.

The work highlights the hidden risk when working on the OHEW while the parallel transmission line is. The case study shows that the current magnitude is high and could jeopardize the human safety (the 615A will be discharged to the ground at the breaking OHEW location.

The proposed mitigations ensure the OHEW section current doesn't contribute to any EPR at the break location. Also the mitigations has no cost or time implication to the proposed works.

VIII. References

[1] Transgrid Network Management Plan 2013-2018
[2] AS/NZS 4853:2012, Standard, electrical hazards on metallic pipelines Applications
[3] Hellany A., Nassereddine M., Nagrial M. and Rizk J. "Transmission mains earthing design: under ground to over head pole transission" International Journal of Electrical, Computer, energy, Electronics and Communication Engineering. Vol. 6, No. 12. pp. 1422-1427. December 2012
[4] IEEE guide to safety in AC substation grounding, 2000' (IEEE, New York, 2000).
[5] AS 2067-2008. Substations and high voltage installations exceeding 1kV AC
[6] Dawalibi F. "Measurements and computations of fault current distribution on overhead transmission lines" IEEE transactions on power apparatus and systems, Vol. PAS-103, No. 3, pp. 553-560. March 1984.
[7] Chang C.N. "Computation of current-division factors and assessment of earth-grid safety at 161/69kV indoor-type and outdoor-type substations" IEE Proc.-Generation Transmission and Distributions. Vol. 152, No. pp. 837-848. 6, November 2005.
[8] Nassereddine M., Rizk J., Hellany A. and Nagrial M. "Relation between transmission lines coupling factor and OHEW length; its impacts on fault current distribution" IET Generation, Transmission & Distribution Journal, Vol. 8, Issue 4, PP.600-608, 2014
[9] Nassereddine M., Rizk J., Hellany A. and Nagrial M. "Transmission Mains pole earth potential rise (EPR) under substation fault" International journal of Emerging Electric Power Systems-Volume 14, Issue 5, pp. 449-507. 2013

Analytical model for the energy harvesting of a spherical sensor from ambient vibrations

D. Diab, F. Lefebvre, G. Nassar, N. Smagin
Department of Opto-Acousto-Electronics
Institute of Electronics, Microelectronics and Nanotechnology UMR CNRS 8520
59313 Valenciennes Cedex 9, France
Daher.Diab@etu.univ-valenciennes.fr

A. Naja, F. El Omar
Laboratory of Physics and Modeling, EDST, Lebanese University
1300 - Tripoli, Lebanon

Abstract— In this work an analytical model for the energy harvesting of an acoustic spherical sensor has been developed in the context to make it autonomous. Our spherical sensor is composed of two half-spheres of Plexiglas and a piezoelectric ring of PZ26 that can be used as exciter or sensor. For the analytical model, the piezoelectric ring was modeled using two primary modes of vibration: thickness and radial. For each mode, the ring is described by an equivalent electromechanical model which connects the mechanical part (forces and velocities) to the electrical part (voltage and current). The proposed paper theoretical model enables building a global electromechanical circuit in order to simulate the total harvested voltage response.

Keywords—energy harvesting; piezoelectric ring; spherical sensor; vibration modes

I. INTRODUCTION

In recent years, much research has been done on the issue of energy autonomy and in particularly on the supply of sensors. The energy harvesting is a theme devoted to the use of the ambient energy (vibration, light, and temperature) present in the environment for powering electronic devices (Sensors, mobile equipment), in a way to extend their operating life and make them completely autonomous.

Ambient energy sources are numerous, and from this, ambient mechanical vibrations are very studied and starting to be used. In a general way, vibratory energy recovery are based on resonant mechanical systems (spring-mass-damper) tuned to the frequency of the source [1]. Several physical phenomena are used for extraction of mechanical energy and its conversion into electrical energy, the piezoelectricity [2], the electrostatic [3] and the electromagnetism [4] can be used as an example of the conversion phenomena. In the context of our work, we focused particularly on piezoelectricity.

A device for recovering energy from ambient vibrations consists essentially of four necessary units: The first unit is a purely mechanical device its role is to capture and optimized mechanical vibrations, The second is an electromechanical device whose purpose is to transform the recovered mechanical energy into electrical energy, The third unit is an electrical device or more precisely an electrical circuit capable of converting the non-exploitable electric energy into exploitable electrical energy, and the last unit is a device for storing the energy before being applied to the sensor to make it electrically autonomous. To improve the power density of the generators and their bandwidth, it is necessary to optimize the four conversion units.

II. SENSOR DESCRIPTION

In this work a spherical device as an acoustic sensor has been developed. This sensor is composed of a closed spherical shell, assembled from two half-spheres made of Plexiglas. Its outer radius is 11 mm and the inner radius is 10 mm. An active element (piezoelectric ring) is sandwiched between the two half spheres [5]. The electronic embarking the sensor application is placed in the hollow portion of the sphere. It ensures the excitation of the piezoelectric ring, the control of the transceiver, the processing and data storage (fig. 1).

Fig. 1. The hollow portion with the embedded electronics.

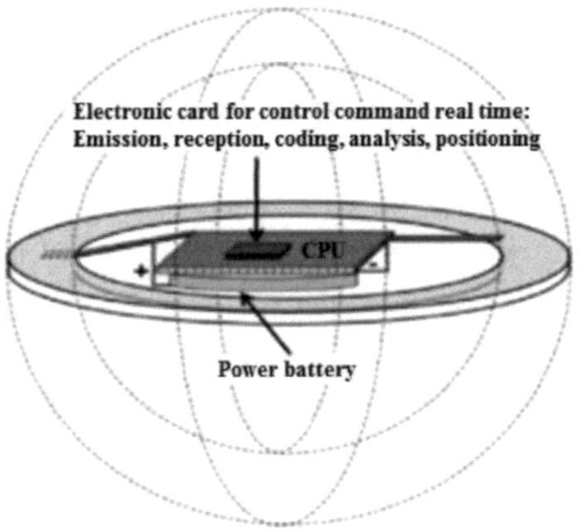

The hollow portion contains a programmable chip (SoC) which ensures the control of the resonator. It includes a central processing unit (CPU), a digital oscillator (DCO), a task scheduler (IT), an encoded ultrasonic generator (PWMG), a power management unit (PMU), an output amplifier (OB), and a battery (BAT).

III. MODELIZATION OF THE PIEZOELECTRIC RING

The harvesting of vibrational energy by piezoelectric transduction has a great interest. Recently, many researchers [6, 7], have worked on the development of systems based on different conversion techniques. In the case of our sensor, the conversion of the vibrational energy into electric energy is achieved by means of a piezoelectric ring undergoing deformations exerted by a sphere excited by external vibrations. For this purpose, the paper models the piezoelectric ring according to its two main modes of vibration: thickness and radial.

A. Thickness mode

Consider an elementary slice of the ring, thickness dz and section A. The displacements u and the forces F are along the longitudinal axis (axis z) as indicated in the figure 2 [8]:

Fig. 2. Elementary slice of the ring in the thickness oscillation

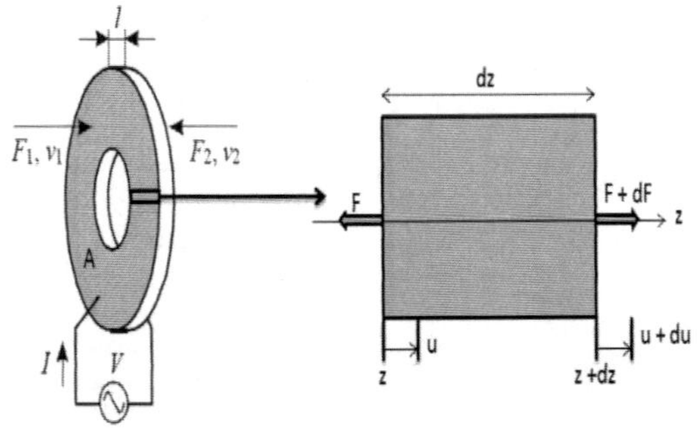

The fundamental principle of dynamics applied to this elementary volume gives [8]:

$$\rho . A . dz \frac{\partial^2 u}{\partial t^2} = \frac{\partial F}{\partial z} dz \tag{1}$$

The constitutive equations of a passive solid material are:

$$T = c_{33} . S \quad T = F/A \quad S = \partial u / \partial x \tag{2}$$

The combination of these equations gives:

$$\rho . \frac{\partial^2 u}{\partial t^2} = c_{33} . \frac{\partial^2 u}{\partial z^2} \tag{3}$$

With ρ is the density of the PZ26 and c_{33} is its Young's modulus.

Assuming the harmonic vibrations and using the complex notation a way of separating the variables $u(z, t) = U(z)e^{j\omega t}$, the equation 3 becomes:

$$\frac{\partial^2 U}{\partial z^2} + \omega^2 . \frac{\rho}{c_{33}} . U = 0 \tag{4}$$

The solutions of this equation are given by:

$$U(z) = \alpha . e^{-jkz} + \beta . e^{jkz} \tag{5}$$

Where $k = \omega . \sqrt{\frac{\rho}{c_{33}}}$ is the wave vector in m-1, α and β are two constants.

The ring has a thickness L, the values of the coefficients α and β depends to the boundary conditions on the faces of the ring and supposing t = 0:

$$\dot{u}(0,0) = v_1 = j\omega (\alpha + \beta) \tag{6}$$

$$\dot{u}(L,0) = v_2 = j\omega (\alpha . e^{-jkL} + \beta . e^{jkL}) \tag{7}$$

This gives:

$$\alpha = \frac{v_2 - v_1 e^{jkL}}{2\omega \sin(kL)} \text{ and } \beta = \frac{v_2 + v_1 e^{-jkL}}{2\omega \sin(kL)}$$

The forces at the ring ends z = 0 and z = L are calculated by:

$$F(0) = F_1 = Z [\frac{(v_1 + v_2)}{j\sin(kL)} + j \tan(\frac{kL}{2}) v_1 \tag{8}$$

$$F(L) = F_2 = Z [\frac{(v_1 + v_2)}{j\sin(kL)} + j \tan(\frac{kL}{2}) v_2] \tag{9}$$

Where Z is the acoustic impedance of the ring; it is expressed as a function of c_{33} and the density ρ by:

$$Z = F/\dot{u} = A . \sqrt{c_{33} . \rho} \tag{10}$$

If take into account the piezoelectricity of the ring, the expression of the constraint has to be modified according to the constitutive equations of the piezoelectric material:

$$T = C^D . S - h_{33} . D \tag{11}$$

With $C^D = c_{33} + \frac{e^2}{\varepsilon_{33}^S}$ and $h_{33} = \frac{e}{\varepsilon_{33}^S}$.

So the two forces are calculated replacing T by T + h.D, which give the following system:

$$F_1 - h_{33} DA = Z [\frac{(v_1 + v_2)}{j\sin(kL)} + j \tan(\frac{kL}{2}) v_1 \tag{12}$$

$$F_2 - h_{33} DA = Z [\frac{(v_1 + v_2)}{j\sin(kL)} + j \tan(\frac{kL}{2}) v_2 \tag{13}$$

Note that the term $h_{33}DA$ represents a force denoted F due to the piezoelectric contribution. Now this electromechanical model can be transposed in an electromechanical equivalent circuit as shown the following figure:

Fig. 3. Electromechanical circuit of the piezoelectric ring in the thickness mode

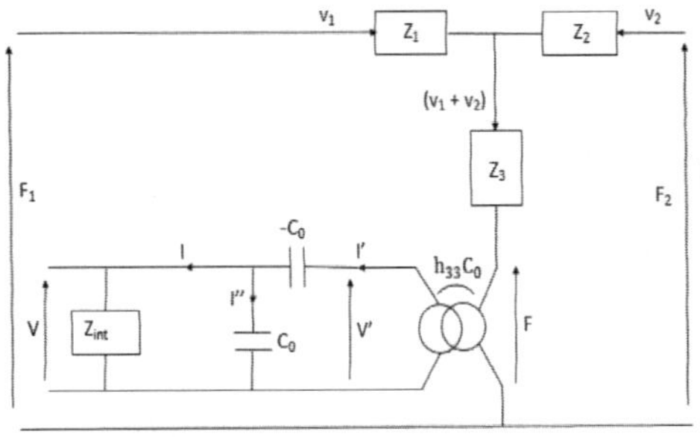

Here, $Z_1 = Z_2 = jZ.\tan(\frac{kL}{2})$, $Z_3 = -j\frac{Z}{\sin(kL)}$ and $C_0 = \frac{\varepsilon_{33}^{S}A}{L}$ the blocked capacity of the ring.

1) Analytical response in voltage

The electromechanical equivalent circuit is formed of two parts connected by a transformer of ratio $h_{33}C_0$, a pure mechanical part designated by the mechanical forces F_1 and F_2 and the displacement velocities v_1 and v_2, an electric part to the output of the transformer designated by the generated electric voltage and current V and I.

This circuit allows us to write:

$$\bar{F} + Z_3\,(v_1 + v_2) = \bar{F}_1 - Z_1 v_1 = \bar{F}_2 - Z_2 v_2 \qquad (14)$$

For a symmetric velocity field compared to the piezoelectric ring plane, we can suppose that $v_1 = v_2$.

This gives:

$$\bar{F}_1 = \bar{F} + (2Z_3 + Z_1)\,v_1 \qquad (15)$$

The voltage and current at the output of the transformer are given by:

$$\bar{V}' = \bar{F}/h_{33}.C_0 \qquad (16)$$

$$I' = h_{33}.C_0(v_1 + v_2) = 2h_{33}C_0v_1 \qquad (17)$$

$$I'' = I' - I = jC_0\omega\bar{V} \text{ and } I = \bar{V}/Z_{int} \qquad (18)$$

From where:

$$\bar{V}\,(1 + \frac{1}{jC_0\omega.Z_{int}}) = \frac{I'}{jC_0\omega} = \frac{2h_{33}}{j\omega}\,v_1 \qquad (19)$$

The equivalent impedance of the electrical part is given by:

$$Z_{elec} = \frac{-1}{jC_0\omega} + \frac{Z_{int}}{1 + jC_0\omega.Z_{int}} \qquad (20)$$

We have $\bar{V}' = Z_{elec}.I'$ which give the relation between the force F and velocity v_1 which is of the following form:

$$\bar{F} = 2h_{33}^2{C_0}^2 Z_{elec}.v_1 \qquad (21)$$

The equation 15 gives:

$$\bar{F}_1 = (2h_{33}^2{C_0}^2.Z_{elec} + 2Z_3 + Z_1)\,v_1 \qquad (22)$$

Therefore the equation 18 becomes of the form:

$$\bar{V}\left(1 + \frac{1}{jC_0\omega.Z_{int}}\right) = \frac{2h_{33}}{j\omega}\,\frac{1}{(2h_{33}^2{C_0}^2.Z_{elec} + 2Z_3 + Z_1)}\,\bar{F}_1 \qquad (23)$$

Where the voltage $\bar{V}$ versus the force $\bar{F}_1$ is:

$$\bar{V} = \frac{2h_{33}C_0.Z_{int}}{(1 + jC_0\omega.Z_{int})(2h_{33}^2{C_0}^2.Z_{elec} + 2Z_3 + Z_1)}\,\bar{F}_1 \qquad (24)$$

B. Radial mode

To obtain the equations of motion, the constitutive equations of piezoelectricity are written in cylindrical coordinates [9]. In this case the variables are the radius r, the angle θ and the axial dimension denoted z (fig. 4). In this figure, L, a and b are respectively thickness, outer and inner radius of the ring, F_{ra}, F_{rb} and v_{ra}, v_{rb} are external radial forces and vibrational velocities at the outer and inner surfaces of the ring.

Fig. 4. The piezoelectric ring in radial mode

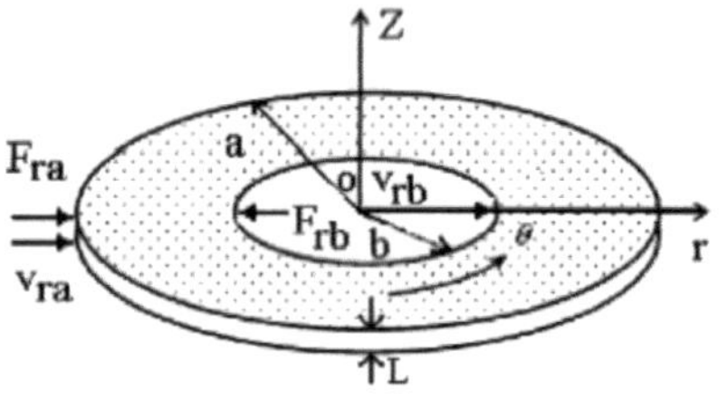

The electromechanical equivalent circuit of the piezoelectric ring in the radial mode is of the following form:

Fig. 5. Electromechanical circuit of the piezoelectric ring in the radial mode

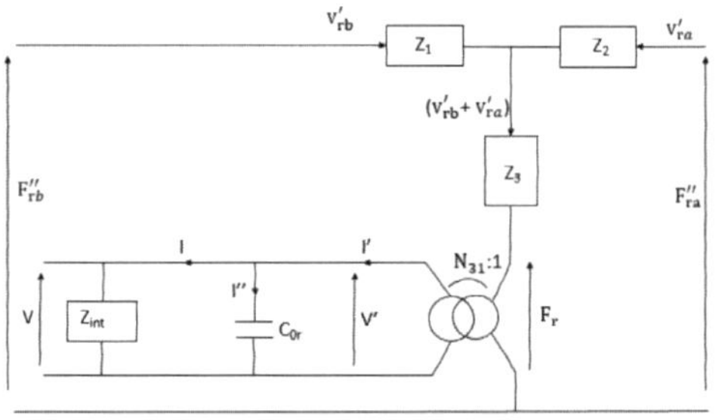

Here, $C_{or} = \frac{\varepsilon_{33}^{T}A}{L}[1 - \frac{2d_{31}^2}{\varepsilon_{33}^{T}(s_{11}^{E} + s_{12}^{E})}]$ is the blocked capacity of the ring in radial mode and $N_{31} = \pi^2\alpha ab.\frac{d_{31}}{s_{11}^{E} + s_{12}^{E}}$ the electromechanical coupling factor of the ring, s_{ij}^{E} is elastic compliance constants measured at constant electric field, d_{31} is piezoelectric strain constant, ε_{33}^{T} is dielectric constant measured at constant stress, $A = \pi(a^2 - b^2)$ is cross-sectional area, $\alpha = \omega/v_r$, $v_r = \sqrt{c_{11}^{E}/\rho}$ are radial wave number and sound speed, $c_{11}^{E} = 1/(S_{11}^{E}(1 - \upsilon_{12}^2))$, with $\upsilon_{12} = -S_{12}^{E}/S_{11}^{E}$ is Poisson ratio.

In figure 5, $F_{ra}'' = (\pi\alpha b/2)\,F_{ra}$, $F_{rb}'' = (\pi\alpha a/2)\,F_{rb}$, $v_{ra}' = (2/\pi\alpha b)v_{ra}$, $v_{rb}' = (2/\pi\alpha a)v_{rb}$ and $F_r = N_{31}V$, with $V = E_zL$, E_z is electric field in the z direction.

The expressions of the three impedances are given by:

$$Z_1 = \frac{\pi^2(\alpha a)^2 Z_{rb}}{4j}\left[\frac{Y_1(\alpha a)J_0(\alpha b) - J_1(\alpha a)Y_0(\alpha b)}{J_1(\alpha a)Y_1(\alpha b) - J_1(\alpha b)Y_1(\alpha a)} + \frac{(1 - \upsilon_{12})}{\alpha b}\right] - j\frac{\pi\alpha a Z_{rb}}{2}\,\frac{1}{J_1(\alpha a)Y_1(\alpha b) - J_1(\alpha b)Y_1(\alpha a)} \qquad (25)$$

$$Z_2 = \frac{\pi^2(\alpha b)^2 Z_{ra}}{4j}\left[\frac{Y_1(\alpha b)J_0(\alpha a) - J_1(\alpha b)Y_0(\alpha a)}{J_1(\alpha a)Y_1(\alpha b) - J_1(\alpha b)Y_1(\alpha a)} - \frac{(1-\upsilon_{12})}{\alpha a}\right] - j\frac{\pi\alpha b Z_{ra}}{2}\frac{1}{J_1(\alpha a)Y_1(\alpha b) - J_1(\alpha b)Y_1(\alpha a)} \quad (26)$$

$$Z_3 = j\frac{\pi\alpha a Z_{rb}}{2}\frac{1}{J_1(\alpha a)Y_1(\alpha b) - J_1(\alpha b)Y_1(\alpha a)} = j\frac{\pi\alpha b Z_{ra}}{2}\frac{1}{J_1(\alpha a)Y_1(\alpha b) - J_1(\alpha b)Y_1(\alpha a)} \quad (27)$$

In these equations, J_0, Y_0, and J_1, Y_1 are Bessel functions of order zero and one respectively, $Z_{ra} = \rho v_r S_a$, $Z_{rb} = \rho v_r S_b$, with $S_a = 2\pi aL$ and $S_b = 2\pi bL$ are outer and inner surfaces of the ring respectively.

The analytical voltage response of this vibration mode calculated from the equivalent circuit is expressed as the following form:

$$\overline{V} = \frac{2N_{31}.Z_{int}}{(1 + jC_{0r}\omega.Z_{int})(2N_{31}^2.Z_{elec} + 2Z_3 + Z_2)} F''_{ra} \quad (28)$$

Here, $Z_{elec} = Z_{int} / (1 + jC_{0r}\omega.Z_{int})$ is the equivalent impedance of the electrical part.

C. *Internal load of energy extraction circuit*

The two equivalent electromechanical circuits for the two modes of vibration are completed with the internal impedance load (Zint) of the processing circuit of the harvested energy. So to consider this charge effect, the power transferred has been modeled depending on the internal impedance of the energy extraction circuit.

The shape of the power (fig. 6) shows that the power reaches a maximum value for $Z_{int} = 24\ K\Omega$. This optimal value of the load will allow a recovery of maximum power of the vibrational radial mode.

Fig. 6. Power modeled depending of the internal impedance

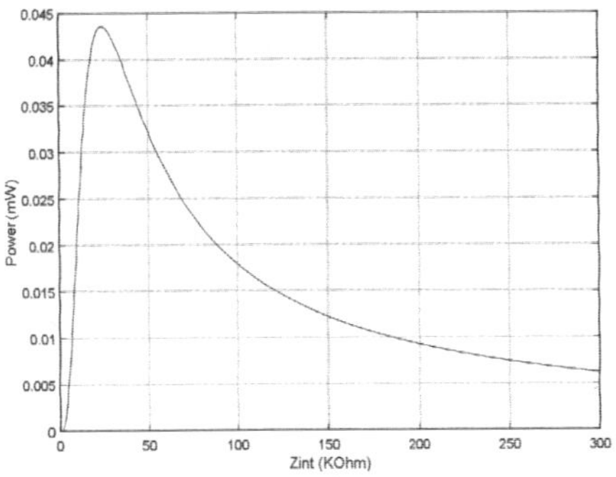

In a same manner, we modeled the power for the longitudinal mode of vibration, the impedance of load obtained are shown in the following table:

TABLE I. INTERNAL IMPEDANCES FOR THE TWO VIBRATIONAL MODES

Mode	Longitudinal	Radial
Z_{int} (KΩ)	28	24

So for the future, we used the internal load impedance of the radial vibration mode (24 KΩ) to determine the total harvested voltage given by the two modes of vibration because the radial mode gives the highest voltage by comparison with the longitudinal mode.

II. ANALYTICAL MODEL OF THE ENERGY HARVESTING

A. *Forces field applied to the ring*

When using our sensor in a flowing fluid, the sphere is deformed under the action of the stresses produced by the fluid; the deformation of the structure creates stresses on the piezoelectric ring. Therefore to calculate the force field applied to the ring due of the fluid flow, a finite element model was built under COMSOL Multiphysics (fig. 7).

Fig. 7. Mesh of the sensor showing the ring zone

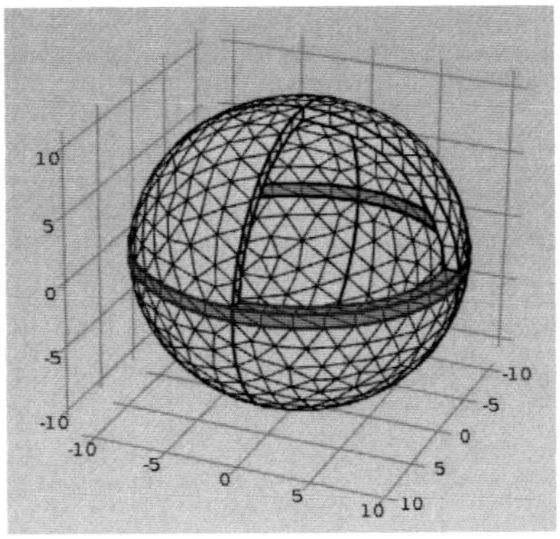

This model makes it easy to combine three approaches: acoustic, mechanical and piezoelectric conversion using three modules "Pressure Acoustics", "Solid Mechanics" and "Electrostatics". So from this model we determined the sound pressure per unit area applied in the plane of the piezoelectric ring which varies depending on the frequency. The integration of this pressure on the entire ring surface gives the force applied on the faces of the piezoelectric ring placed at the junction of the two half spheres (fig. 8).

Fig. 8. Force applied in the ring plane

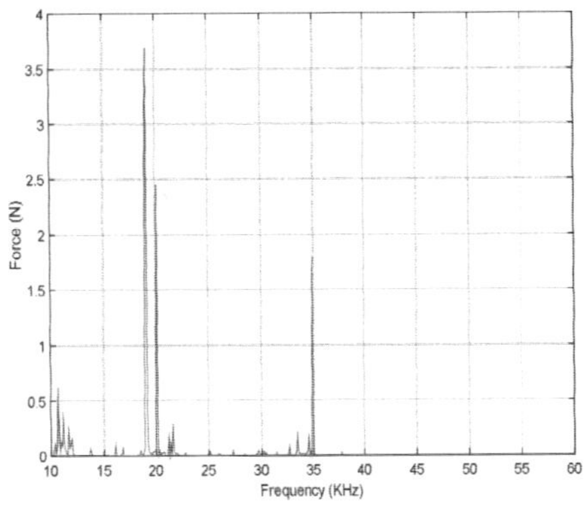

This figure shows the shape of the force field applied to the ring for the frequency band between 10 and 60 KHz. This force is then applied as data input to the two vibrational models of the piezoelectric ring.

B. Proposed analytical model

The objective of our theoretical modeling is to assemble the two electromechanical circuits corresponding to the two modes of vibration in a way to simulate the total recovered energy treated by a specific electronics and stored in a capacitor (fig. 9). The entire modeling must be subsequently made under Spice software which is able to simulate electronic circuits.

Fig. 9. Theoretical model of the two-mode simulation

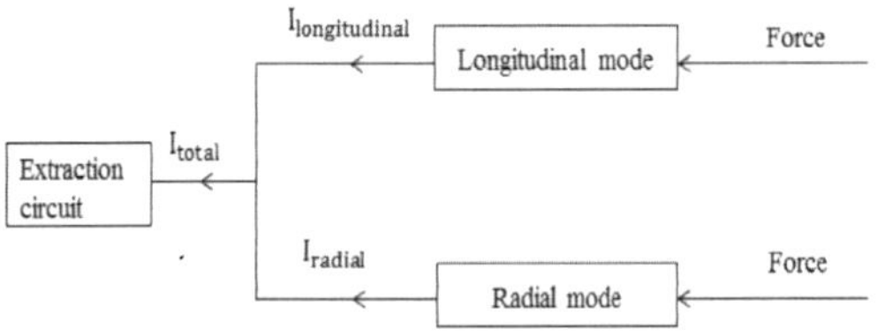

This model has a pure mechanical input designated by the mechanical forces exerted on the piezoelectric ring of our spherical sensor; these forces are described in IV.A. This analytical model given as output the electrical voltage and current generated by the two electromechanical diagrams. So to determine the total recovered voltage, it is necessary to model the two modes of vibration using the same internal load impedance, and in this case we have proposed to use the impedance of the radial mode (24 KΩ) because this last generates the highest voltage. So the total voltage produced by our theoretical model is only the sum of the two voltages given by the two modes of vibration.

$$V_{total} = V_{longitudinal} + V_{radial} \quad (29)$$

A Matlab script has been written to implement the analytical equation 29, and calculate the total voltage response of our proposed model in the frequency range from 10 kHz up to 60 kHz. The obtained results are shown in the following figure:

Fig. 10. Analytical voltage response of our theoretical model

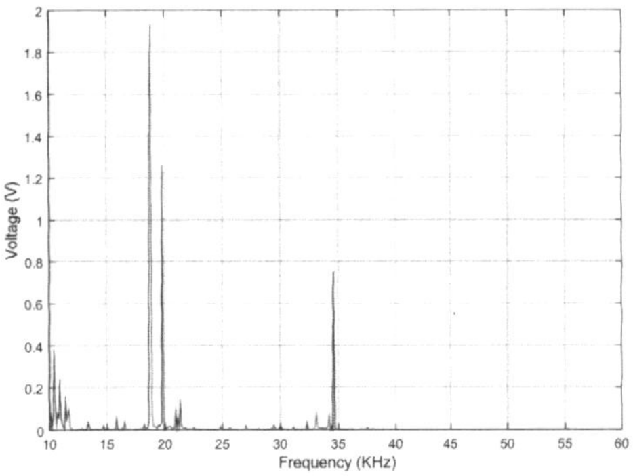

III. CONCLUSION

In this study, we sought to develop an analytical model for the recovery of vibrational energy of a spherical sensor. This sensor consists of a piezoelectric ring sandwiched between two half-spheres. This model must produce the electromechanical conversion potential of such a sensor according to its two main modes of vibration. For each mode, an equivalent circuit has been established to be integrated with the Spice simulation model of the energy processing circuit. A finite element model was constructed under COMSOL Multiphysics to determine the input force for the two vibrations modes. The experimental validation of this approach is currently underway.

REFERENCES

[1] Cottone, F., Introduction to vibration energy harvesting. NiPS Energy Harvesting Summer School, 2011.J. Clerk Maxwell, A Treatise on Electricity and Magnetism, 3rd ed., vol. 2. Oxford: Clarendon, 1892, pp.68–73.

[2] Ajitsaria, J., Modeling and analysis of a bimorph piezoelectric cantilever beam for voltage generation. April 2007.K. Elissa, "Title of paper if known," unpublished.

[3] S. Meninger, Vibration to electric energy conversion. IEEE Transactions on Very Large Scale Integration (VLSI), February 2001.

[4] Glynne-Jones, P., An electromagnetic, vibration powered generator for intelligent sensor systems. February 2004.

[5] Medjdoub, A., Study and conception of a spherical acoustic sensor, miniaturized, encoded and autonomous. may 2016.

[6] Sainthuile, T., Energy Harvesting Process Modelling of an Aeronautical Structural Health Monitoring System Using a Bond-Graph Approach. International Journal of Aerospace Sciences, 2012.

[7] Erturk, A., An experimentally validated bimorph cantilever model for piezoelectric energy harvesting from base excitations. IOPscience, 13 January 2009.

[8] Lin, S., Study on the sandwich piezoelectric ceramic ultrasonic transducer in thickness vibration. IOP science 8 January 2008.

[9] Lin, S., Electro-mechanical equivalent circuit of a piezoelectric ceramic thin circular ring in radial vibration. July 2006.

APPENDIX

TABLE II. PLEXIGLAS PARAMETERS USED IN THE SIMULATION OF THE SPHERICAL SHELL

Elasticity module E (MPa)	Poisson coefficient ν	Density ρ (kg/m³)
3300	0.39	1190

TABLE III. PZ26 PARAMETERS USED IN THE SIMULATION OF THE PIEZOELECTRIC RING

ε_0(F/m)	$\varepsilon^s_{33}/\varepsilon_0$	$\varepsilon^T_{33}/\varepsilon_0$	ρ(kg/m³)	h_{33}(V/m)	e_{33}(C/m²)
$8.8\ 10^{-12}$	700	1300	7700	$23.7\ 10^{8}$	14.7
d_{31}(C/N)	c^D_{33}(N/m²)	c^E_{11}(N/m²)	s^D_{33}(m²/N)	s^E_{11}(m²/N)	s^E_{12}(m²/N)
$-130\ 10^{-12}$	$9.56\ 10^{10}$	$7.7\ 10^{10}$	$11\ 10^{-12}$	$13\ 10^{-12}$	$-4.35\ 10^{-12}$

Influence of Processing Parameters on the Tensile Strength of PA-W/mLLDPE Composite Films Produced by Blown Film Extrusion

Nancy Zgheib [(1)], *Doumit Tahan* [(1)], *Sylvain Seif* [(2)], *Nemr El Hajj* [(1)]
[(1)] *Department of Chemical and Mechanical Engineering, Faculty of Engineering, Holy-Spirit University of Kaslik, P.O. Box 446, Jounieh, Lebanon*
[(2)] *PACT R&D Center, INDEVCO Group, Zouk Mosbeh, Lebanon*

Abstract — **Extrusion blown films from Polyamide waste/metallocene linear low-density polyethylene (PA-W/mLLDPE) blend based matrix system and calcium carbonates particles were successfully produced and characterized. Also, the effect of the extruder temperature used during twin-screw compounding, the coupling agent loading and PA loading on the mechanical properties of the obtained films were studied. The key aspect of the present study was in investigating the optimal combination of factors by using Design of Experiment (DOE) approach. The responses measured were the tensile strength in the machine direction (MD) and the tensile strength in the transverse direction (TD). Among the selected processing parameters, the coupling agent and the interaction between the coupling agent and the PA loading have the most important influence on the tensile strength.**

Keywords — PA-W/mLLDPE composites, blown film extrusion, Design of Experiment (DEO), tensile test

I. Introduction

Nowadays, plastic recycling is becoming an important challenge. The volumes of plastic wastes are increasing day by day, requiring effective recycling processes of different kinds of materials in a cost-efficient way [1]. Polyamide (PA) or nylon is one of the most common plastics used in food packaging. It is also used for medical and pharmaceutical packaging. Up to now, recycling of PA wastes (PA-W) is not well elaborated and few studies are available in the literature [2]. Several advances have been achieved in the plastic recycling technology; depolymerization by chemical recycling is a way to replay original monomers [3]. However, this method is difficult to achieve due to the complexity of systems and high cost of operation. The burning of polymers to recover energy is used too. In this case, environmental benefits are debatable [4]. Mechanical recycling is an interesting alternative which can combine technical viability, acceptable costs and environmental benefits. The presence of impurities and the degradation of the material are the most important obstacles for this method of polymer recycling [5, 6]. In this work mechanical recycling was used to prepare polymer blends. PA-W/metallocene linear low-density polyethylene (mLLDPE) composite blends were produced using high-shear mixing twin screw extruder and the films were manufactured using a blown film extrusion process.

Within the scope of this study, a design of experiment (DOE) approach was investigated to determine the influence of many input variables on the output of the process used. Therefore, a series of structured tests were planned and the effects of the changes are then evaluated. A DOE study was made by Halim and al. [7] to investigate the influence of processing parameters on the mechanical properties of polyamide 11 composites. From this study, it was found that the nanoclay loading and the maximum barrel temperature are the significant variables that affect the mechanical properties of polyamide 11 composites. Furthermore, the influence of processing variables on the impact properties of the poly(butylene succinate) PBS/poly(butylene adipate-co-terephthalate) PBAT/miscanthus biocomposites using a full factorial experimental design was studied by Mathuraj and al. [8]. The author demonstrates that the fiber length has the most important influence on the impact strength of the biocomposites. Another study by Johnson and al. [9, 10] showed that processing temperature affect largely the impact performance of Mater-Bi/miscanthus composites.

The aim of the present work is to optimize the fabrication of plastic films using PA-W/mLLDPE matrix with calcium carbonate as the reinforcing agent and varying several processing conditions (extruder temperature, coupling agent loading and PA loading). A full factorial design and analysis of variance (ANOVA) method was considered to analyze the effect of these parameters on the mechanical properties of the produced films. The effect of calcium carbonate loading level was not investigated in this work due to the fact that this loading level had already been optimized in a previous study. Besides, due to confidentiality agreements, applications of these plastic films are kept confidential.

II. Experimental Procedure

A. Materials

The thermoplastic waste used in this work was PA-W from cast film extrusion obtained from INDEVCO, Lebanon. The PA-W is a copolymer 6/66 grade of intermediate viscosity. The molecular weight of the $C_{18}H_{37}N_3O_5$ monomer is 375.5 g/mol. The crystal structure of this PA-W is a shish-kebab type. The PE is a mLLDPE obtained from Exxon Mobil Chemical. The coupling agent is an anhydride-modified LLDPE resin obtained from DuPont Packaging & Industrial

Polymers. Calcium carbonates were obtained from Jordan Carbonate Company.

B. Full factorial design

A full factorial design was chosen in order to identify the influence of each variable as well as the interactions between the variables on the response. Using this statistical approach, there are 2k experiments, where 'k' is the number of variables. In this study, the processing conditions being evaluated are coupling agent loading, PA loading and extruder temperature. As shown in TABLE I, two levels were assigned for each factor. MINITAB software was employed to make the experimental design and to analyze the obtained results through statistical plots at a confidence level of 95%. For three independent variables there are eight experiments, each corresponding to a particular combination of factors as presented in TABLE II.

TABLE I. INDEPENDENT VARIABLES AND THEIR RESPECTED LEVELS

Variables	Name	Lower level	Higher level
1	Extruder temperature (°C)	230	260
2	Coupling agent loading (%)	2.5	5
3	PA loading (%)	15	23.5

C. Melt compounding and blown film extrusion

The production of the films is a two-stage process. The first stage is to produce a masterbatch by melt processing using a co-rotating twin screw extruder. At this stage, PA-W pellets are mixed with mLLDPE in the presence of the coupling agent and calcium carbonate filler to produce homogenized pellets or masterbatches. Before using the PA-W pellets, they were subjected to an overnight drying. The second stage of the film production is the processing of the masterbatches in a blown film extrusion. The film is pulled in the transverse direction (TD) to reach the ultimate diameter of the bubble. At the same time, it is being pulled in the machine direction (MD) by the movement of the downstream equipment. All twin screw compounding processes were carried out on a Leistritz Micro ZSE 27 HP twin screw extruder with a 27 mm screw diameter and a 48/1 length to diameter (L/D) ratio.

D. Mechanical testing

Tensile testing of plastic films was performed on an LLOYD universal testing machine using a 1 kN load cell with a crosshead speed of 500 mm/min in accordance with ASTM D882–12 at ambient temperature. A minimum of 5 test specimens were tested from each batch where the average data will be presented later in this paper. Tensile strength, tear resistance and dart drop impact tests were applied to the product to determine its mechanical properties. In this paper, we will present only the tensile test results.

III. RESULTS AND DISCUSSION

A DOE analysis was conducted to find out the effect of three factors on tensile strength of different blends. The objective of this study was to point out the most advantageous combination of factors (coupling agent loading, PA loading and extruder temperature). The results for the considered

TABLE II. DOE VARIABLES AND PA-W/MLLDPE COMPOSITE FORMULATIONS

Experiment	Extruder temperature (°C)	Coupling agent loading (%)	PA loading (%)	Code
1	230	5	15	230T/5C/15P
2	230	2.5	23.5	230T/2.5C/23.5P
3	260	2.5	23.5	260T/2.5C/23.5P
4	260	5	15	260T/5C/15P
5	230	2.5	15	230T/2.5C/15P
6	260	2.5	15	260T/2.5C/15P
7	230	5	23.5	230T/5C/23.5P
8	260	5	23.5	260T/5C/23.5P

TABLE III. MECHANICAL RESULTS FOR EACH BLEND

Run	Code	Tensile Strength in MD (MPa)	Tensile Strength in TD (MPa)
1	230T/5C/15P	20.5±0.97	16±1.6
2	230T/2.5C/23.5P	5.8±0.52	5.3±0.17
3	260T/2.5C/23.5P	12.6±2.18	9.1±0.67
4	260T/5C/15P	18.1±2.61	16±1.31
5	230T/2.5C/15P	14.6±1.35	13.3±2.01
6	260T/2.5C/15P	16.6±1.02	16.2±0.95
7	230T/5C/23.5P	24.9±3.12	19.9±1.71
8	260T/5C/23.5P	24.2±1.9	20.3±0.47

responses (tensile strength in MD and TD directions) are shown in TABLE III. When comparing the results obtained for all the blends, it can be seen that the increase in the coupling agent loading from 2.5% to 5% increases the tensile stress in the machine direction and in the transverse direction for at least 40% for both configurations. Also, we can notice that an interaction exist between the coupling agent loading, the PA loading and the extruder temperature. This interaction will be studied further by the interaction plots and the analysis of variance (ANOVA).

A. Pareto chart and Normal plot of the effects

In order to visibly prove the impact of the independent variables on the mechanical properties of the different blends, the Pareto chart of effects is used. This chart shows the actual magnitude of each effect. Any factor that surpasses the reference line identified in red in the following chart is possibly important. Fig.1 shows the significant factors influencing the tensile strength in MD direction of the PA-W/mLLDPE composite with a confidence level (α) of 0.05. According to the Pareto chart, the coupling agent B had an important influence on the tensile strength of the produced films since the standardized effect value is greater than the reference line value 8.56.

Based on the normal plot of the effects, the points which appear far from the fitted line express a significant effect on the tensile strength. The points which are close to the fitted line show insignificant effects on the tensile strength. On the other hand, if the impact of the factor is negative, the concerning point will fall on the left side of the blue line. In Fig.1, it can be seen that the coupling agent is a significant factor since it is represented as square symbol and appears far away from the fitted line.

B. Main effects plot

It is possible to assess the effect of each parameter on the obtained results by using main effects plots. A main effect exists when the line connecting the mean values of the parameter is not horizontal. The sharper the slope of the line, the greater is the importance of the effect. On the other hand, if the line is parallel to the x-axis, thus no main effect is present. Fig.2 shows the main effects plot for the tensile strength in MD and TD directions of the resulting composite. Noticeably, coupling agent loading used during compounding has the greatest effect on the mechanical properties of the blends which is illustrated with a strong line slope. On the other hand, the extruder temperature and PA loading have a slight effect on the tensile strength in TD and MD directions. Also, PA loading shows a negative effect on the response when varying its levels.

C. Interaction Plot and Analysis of Variance (ANOVA)

In order to evaluate the interaction of two factors or more on the obtained results and examine their strength, the interaction plot is utilized. If the lines drawn by the software are not parallel, there is possibility for an interaction between the concerned factors and vice-versa.

ANOVA can be used to study the effects of factors with respect to response. The individual and interaction factors influence on the mechanical properties are estimated by using the sum of square (SS), mean square (MS), F-test statistics and P-values. In order to evaluate the effect of the chosen factors on the factorial design experiment, a 95% confidence interval was selected with an alpha level (α) = 0.05. Usually, if the P-value is less than 0.05, the considered factor has significantly affected the experiment response.

a)

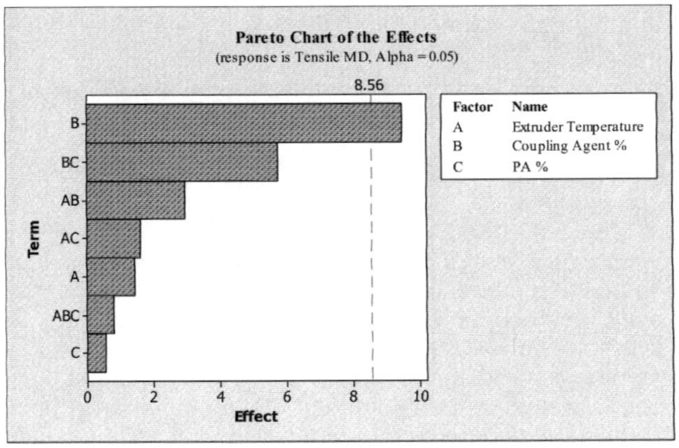

b)

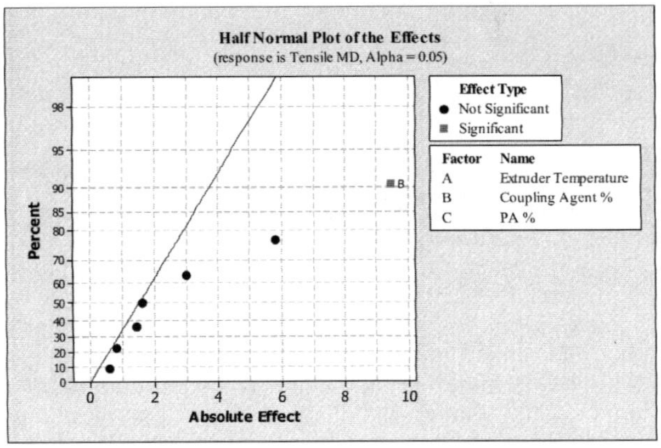

Fig. 1. Represents a) the Pareto chart and b) the Half normal plot of the effects for the tensile strengh in MD direction of the PA-W/mLLDPE composite

a)

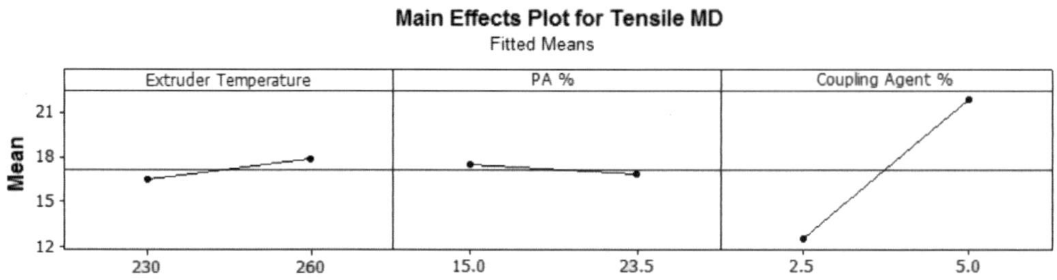

b)

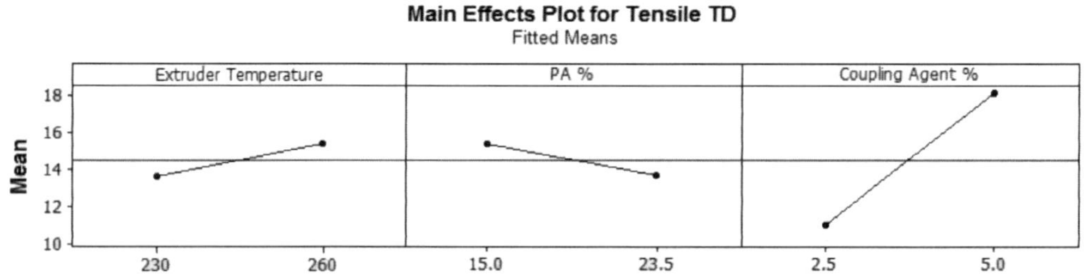

Fig. 2. Represents the main effects plot for tensile strength in a) MD and b) TD directions of the PA-W/mLLDPE composite

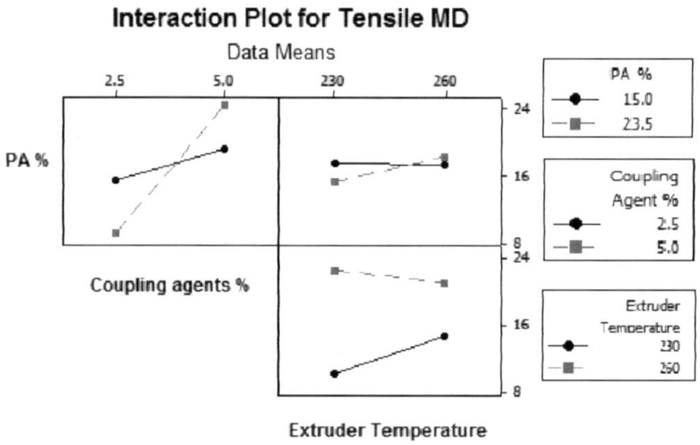

Fig. 3. Interaction plot for tensile strength in MD direction of the PA-W/mLLDPE composite

Fig.3 represents the statistically significant binary interaction effect between the selected variables on the tensile strength in MD direction. Clearly, the PA loading/coupling agent loading, PA loading/extruder temperature and coupling agent loading/extruder temperature have an interaction. Out of the selected variables the combination PA loading/coupling agent loading had the most significant effect on tensile strength while the combination coupling agent loading/extruder temperature had the least significant effect. These results are confirmed by the analysis of variance illustrated in TABLE IV when the P-values between the different combinations are compared. Fig.4 illustrates the interaction plots for the tensile strength in TD direction of the composite. The three combinations show a possible interaction. It can be noticed that the most significant interaction among the different parameters is the combination PA loading/coupling agent loading. The analysis of variance results illustrated in TABLE V shows that the three chosen variables for this study have an impact on the tensile strength in the TD direction. PA loading, coupling agent loading and extruder temperature have respectively a P-value of 0.041, 0.01 and 0.04. According to these results the coupling agent loading has the lowest P-value; consequently it is the most influencing factor for this property. The combination PA loading/coupling agent loading and coupling agent loading/extruder temperature have also an effect on the tensile strength in TD direction.

TABLE IV. ANALYSIS OF VARIANCE FOR TENSILE STRENGTH IN MD DIRECTION OF THE PA-W/MLLDPE COMPOSITE

Source	DF	Sum of squares (SS)	Mean squares (MS)	F	P
PA %	1	0.657	0.657	0.52	0.601
Coupling agent %	1	180.273	180.273	143.64	0.053
Extruder temperature	1	4.042	4.042	3.22	0.324
PA %*Coupling agent %	1	66.809	66.809	53.23	0.087
PA %*Extruder temperature	1	5.184	5.184	4.13	0.291
Coupling agent % * Extruder temperature	1	17.754	17.754	14.15	0.165
Error	1	1.255	1.255		
Total	7	275.974			

S = 1.12027 R-SQ = 99.55% R-SQ(ADJ) = 96.82%

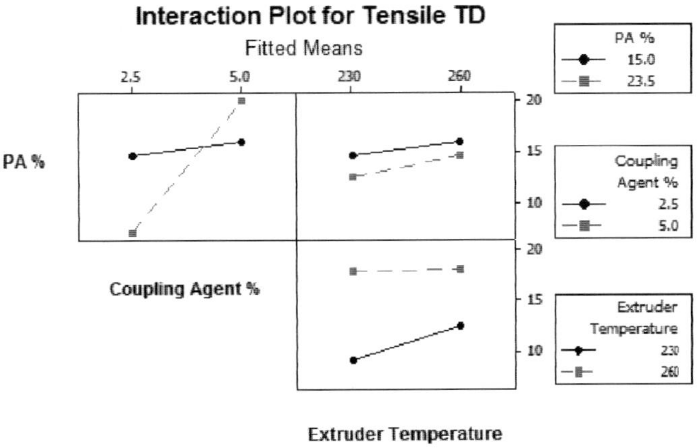

Fig. 4. Interaction plot for Tensile Strength in TD direction of the PA-W/mLLDPE composite

TABLE V. ANALYSIS OF VARIANCE FOR TENSILE STRENGTH IN TD DIRECTION OF THE PA-W/MLLDPE COMPOSITE

Source	DF	Sum of squares (SS)	Mean squares (MS)	F	P
PA %	1	6.016	6.016	239.66	0.041
Coupling Agent %	1	100.74	100.74	4013.46	0.01
Extruder Temperature	1	6.431	6.431	256.19	0.04
PA %*Coupling agent %	1	67.507	67.507	2689.46	0.012
PA %*Extruder temperature	1	0.225	0.225	8.96	0.205
Coupling agent % * Extruder Temperature	1	5.097	5.097	203.07	0.045
Error	1	0.025	0.025		
Total	7	186.041			

S = 0.158432 R-Sq = 99.99% R-Sq (adj) = 99.91%

The low P-value (0.012) for the interaction of PA loading/coupling agent loading in addition to the interaction plot, confirm that this combination is the worth mentioning one.

IV. CONCLUSION

PA-W/mLLDPE composite films have been successfully produced and characterized in this study. The influence of the independent processing variables on the tensile strength of the produced films in machine direction and transverse direction has been investigated by 2^3 full factorial design of experiment. The statistically significant main and interaction variables were analyzed at a 95% confidence level. According to main effect plot, Pareto plot, and half normal plot, the coupling agent loading plays an important role on the tensile strength of these films. On the other hand, the interaction plot and the analysis of variance show that the binary interaction between PA loading/coupling agent loading has the most significant response among the selected variable combinations. This result proves that the maleic anhydride is a suitable coupling agent for compatibilizing normally incompatible polymers like PA and mLLDPE. For future work, we can choose more number of parameters such as the particle size of calcium carbonate filler or the screw speed in order to get the optimum mechanical properties.

ACKNOWLEDGMENT

This work has been funded by INDEVCO PACT (Polymer Application Center for Technology) R&D Center. We give special thanks to Dr. Sylvain Seif and Dr. Johnny Najem of PACT for their support and fruitful scientific discussions.

REFERENCES

[1] B. Acevedo, A.M. Fernández, C. Barriocanal, Identification of polymers in waste tyre reinforcing fiber by thermal analysis and pyrolysis, Journal of Analytical and Applied Pyrolysis, 111 (2015) 224–232.

[2] E. Silva, Recycled polyamides, a literature review and research opportunities, College of Textiles, NCSU, 2008: 2-7, 14.

[3] H. Nishida, Development of materials and technologies for control of polymer recycling, Polym. J. 43 (2011) 435–447.

[4] S.M. Al-Salem, P. Lettieri, J. Baeyens, The valorization of plastic solid waste (PSW) by primary to quaternary routes: from re-use to energy and chemicals, Prog. Energy Combust, 36 (2010) 103–129.

[5] R. Laryea-Goldsmith, J. Oakey, N.J. Simms, Gaseous emissions during concurrent combustion of biomass and no-recyclable municipal solid waste, Chem. Cent. J. 5 (2011) 2–10

[6] C T. Ferreira, C. A.B. Perez, D. Hirayama, C.Saron: Recycling of polyamide (PA) from scrap tires as composites and blends, Journal of Environmental Chemical Engineering, 1 (2013) 762–767.

[7] K. A. Abdul Halim, J. B. Farrell, J.E. Kennedy, Optimisation of a twin screw extrusion process for enhanced short-term mechanical properties of polyamide 11 nanocomposites using Design of Experiment (DOE) approach, Conference paper, June 2012.

[8] R. Muthuraj, M. Misra, F. Defersha, A. Mohanty, Influence of processing parameters on the impact strength of biocomposites: A statistical approach, Composites: Part A 83 (2016) 120–129.

[9] R.M. Johnson, N. Tucker, S. Barnes, Impact performance of miscanthus/NovamontMater-Bi_ biocomposites, Polym Test 2003, 22 (2): 209–15.

[10] M. Johnson, N. Tucker, S. Barnes, K. Kirwan, Improvement of the impact performance of a starch based biopolymer via the incorporation of Miscanthus giganteus fibres, Ind Crops Prod 2005, 22 (3):175–86.

Influence of synthesis parameters of mesocellular silica foams doped by nickel on methane reforming by CO_2

O.Daoura, M. Boutros.
Laboratoire Chimie-physique des matériaux (LCPM/PR2N)
Lebanese University, Faculty of Sciences II (UL/FSII)
Beirut, Lebanon

O.Daoura, F.Launay, P. Massiani.
Laboratoire de Réactivité de Surface (LRS)
Université Pierre et Marie Curie (UPMC)
Paris, France

N. EL Hassan
Chemical engineering department
University Of Balamand (UOB)
Koura, Lebanon

***Abstract*— New catalysts based on Ni(0) dispersed onto mesocellular silica foams (MCF) were prepared for Dry Reforming of Methane (DRM). Different synthesis methods of MCF supports (with and without using NH_4F or *n*-butanol) were tested in order to study the influence of the textural properties of the support on the catalytic performances of the catalysts. In all cases, nickel was incorporated using the double solvents method. The resulting calcined materials were characterized by N_2 sorption, X-Ray diffraction (XRD) and transmission electron microscopy. Their reducibility was tested by temperature-programmed reduction (TPR). Interestingly, particularly large specific surfaces, pores diameters and pores volumes were observed when NH_4F was added to the synthesis gel (MCF(N) materials). The corresponding Ni-MCF(N) catalysts were shown to be the most attractive among other prepared samples with respect to their performances in DRM.**

Keywords — Methane reforming ; Nickel ; Nanoparticles ; Mesocellular foams.

I. INTRODUCTION

In 1999, siliceous mesocellular foams (MCF) were prepared for the first time by combining self-assembly processes through a $(S^0H^+)(X^-I^+)$ pathway used for SBA-15 synthesis and pore swelling with organic molecules [1]. Preferentially solubilized in the core of micelles, these molecules lead, as expected, to an increase of the pore volume but also to a modification of the porosity organization [2]. Since, resulting MCF with large spherical cell sizes up to 40 nm, 3D interconnected pore system and large pore volumes [3] have shown great potential as supports for drug molecule release [1] or for the design of heterogeneous catalysts through the incorporation of enzymes/proteins [4], metal nanoparticles [5] to name only a few. Up to now, various synthesis methods of MCF have been reported in the literature. Among them, the challenge is to obtain complex mesophases using a four-component microemulsion (surfactant/co surfactant/oil/water) as a template [6]. Several examples of co-surfactant / oil combinations have been described in the literature. Among the molecules added, n-butanol (*n*-BuOH) is known to act both as a co-solvent and co-surfactant in block copolymer-water systems. In other terms, *n*-BuOH stabilizes the hydrophilic-hydrophobic interfaces acting on the micellar surface curvature and on the total volume of the hydrophobic chains of the block copolymer [7].

In the present contribution, MCF doped by 5 wt.% of nickel was used for the first time as a catalyst for Dry Reforming of Methane (DRM). Studies on DRM present both scientific and industrial interest. This reaction consists in transforming two abundantly available greenhouse gases (CH_4 and CO_2) to produce industrially important syngas (a mixture of CO and H_2). The latter is usable in the synthesis of a broad range of higher alkanes through Fischer–Tropsch synthesis [8]. However, DRM is accompanied by secondary reactions such as Methane Decomposition (MD), Reverse Water Gas Shift (RWGS) and Boudouard reaction (BD) [9]. To limit them, to favor the endothermic DRM reaction and to reach high equilibrium conversion of CH_4 and CO_2 to H_2 and CO, operating temperatures of 800-1000°C are usually required. This also minimizes side reactions [8]. For these reasons, inexpensive catalysts affording the expected CO/H_2 molar ratio are highly desired in order to lower the operating conditions.

In this work, four types of MCF materials with different textural properties were synthetized with the aim to select the most convenient supports for DRM catalysts preparation. The latter were obtained after deposition of 5 wt.% Ni using the double solvents method [10]. The porosity and the structure of the solids as well as the nickel dispersion were analyzed. The catalysts were tested in dry reforming of methane after reduction. Their catalytic performances in terms of conversions, products selectivity and stability were analyzed and compared to thermodynamic curves simulated by the HSC program considering or not carbon formation and similar reaction conditions.

II. Experimental part

A. Catalysts preparation

A typical synthesis of MCF, as described by Widyaningrum *et al* [11], was used first. It consisted in fully dissolving 4.0 g of Pluronic (P123) triblock copolymer ($EO_{20}PO_{70}EO_{20}$) in a mixture of HCl (20 mL, 37%) and distilled water (120 mL) at 40°C, then adding dropwise 4.0 g of trimethylbenzene (TMB) as swelling agent, under vigorous stirring. To form a microemulsion, the suspension was stirred for 2 h then tetraethylorthosilicate (TEOS, 9 mL) was added dropwise and stirring was continued for 20 h at the same temperature. The solution was next transferred in an autoclave and submitted to a hydrothermal treatment at 100°C for 24 h. The resulting suspension was finally filtered and the solid was washed, dried and calcined under air flow at 550°C for 5 h (heating rate 1°C.min^{-1}). The obtained solid was named MCF. Similar syntheses conditions were used to prepare sample MCF(N), except for the addition of 48 mg of NH_4F done just before the hydrothermal treatment. Also, two other MCF silica foam materials, i.e. MCF(B1) and MCF(B2), were prepared using n-butanol as a co-surfactant following the synthesis procedure proposed by X. Zhou et al [7]. The considered molar *n*-BuOH / TMB ratio was equal to 2, which corresponded to the addition of 2 g of TMB and 2.4 g of butanol for MCF(B1), and of 4 and 4.6 g, respectively, in the case of MCF (B2). Both, surfactant and co-surfactant, were added together.

Ni-MCF, Ni-MCF(N) and Ni-MCF(B1) materials were obtained by the impregnation of the supports with a Ni^{2+} containing solution using the double solvents method. The used nickel precursor was $Ni(NO_3)_2.6H_2O$ and the theoretical loading of Ni added was 5 wt%. In details, the support (1.0 g) was suspended in 20 mL of n-hexane. After that, the appropriate amount of nickel precursor dissolved in a volume of water set equal to the silica pore volume (as determined by N_2 physisorption) was added dropwise. The impregnated solids were dried 3 days at room temperature then overnight at 60 °C then calcined at 550 °C under air for 2 h.

B. Catalysts characterisations

The catalysts were characterized in terms of textural, structural, reducibility and morphological properties. By recording N_2 adsorption-desorption isotherms on an ASAP 2020 (Micromeritics) apparatus the textural properties of solids were obtained. The samples were degassed, prior to experiments, under vacum for 2 h at 250°C. The specific surface area values were obtained by using the BET equation. The pore diameters and specific pore volumes were calculated according to the BJH model. X-Ray Diffraction (XRD) data were recorded on a diffractometer X'Pert3 Powder (PANalytical, Netherlands) using CuKα radiation. The average particle sizes were calculated from the Scherer equation: $D=K\lambda/\beta\cos\theta$, where K is a constant (K=0.9), λ = 1.5405 Å, β is the full width at half maximum (FWHM) of the diffraction peaks and θ is the peak position. Temperature-Programmed Reduction (TPR) done in a TPD/R/O instrument (Autochem 2920 Micromeritics apparatus) was used to inform on the nickel phase reducibility and on the interaction of the metal with the support. The catalysts (50 mg) were placed on a quartz wool in a U-shaped quartz sample tube and heated from room temperature to 800°C at a rate of 10°C.min^{-1} under a gaseous mixture flow (25 mL.min^{-1}) of 5% H_2 diluted in Ar. Before reaching the thermal conductivity detector (TCD), the gas flow was passed through an ice bath to remove any water product formed during reduction. Transmission Electron Microscopy (TEM) analyses were performed on a JEOL-2010 (LaB) microscope operating at 200 kV. Ethanol was used to suspend the powdered samples. The resulting mixtures were deposited on a copper grid coated with a porous carbon film, then ethanol was left to evaporate by drying in room atmosphere.

C. Catalytic measurements

The catalysts were tested in a PID ENG & TECH Microactivity Effi Reactor. They were loaded into the reactor and reduced *in situ* at 650°C for 2 h (rate 10°C.min^{-1}) in a 5% H_2/Ar flow (30 mL.min^{-1}) to form reduced nickel known to be the active phase in DRM. After this activation treatment, the temperature was cooled down to 200°C and dry reforming of methane was conducted under atmospheric pressure with CH_4/CO_2= 1:1 (diluted with Ar) and a total GHSV=36 L.g^{-1}.h^{-1} (100 mg of catalyst were used for each test). Good reproducibility of catalytic results under these conditions was obtained. For activity measurements, the reaction temperature was increased from 200°C (5°C.min^{-1}) up to 750°C followed by a decrease of temperature till 650°C which was maintained for 4 h. The gaseous products were analyzed online by a micro GC Agilent 490 mGC. The conversions of methane (1) and of carbon dioxide (2), as well as the H_2/CO ratios (3) were calculated as follows:

$$CH_4 \text{ conversion } (\%) = \frac{CH_{4(in)} - CH_{4(out)}}{CH_{4(in)}} \times 100 \quad (1)$$

$$CO_2 \text{ conversion } (\%) = \frac{CO_{2(in)} - CO_{2(out)}}{CO_{2(in)}} \times 100 \quad (2)$$

$$H_2/CO = \frac{H_{2(out)}}{CO_{(out)}} \quad (3)$$

where $CH_{4(in)}$ and $CH_{4(out)}$ (resp. $CO_{2(in)}$ and $CO_{2\,(out)}$) are the initial and measured CH_4 (resp. CO_2) concentrations, respectively, whereas $H_{2(out)}$ and $CO_{(out)}$ are the measured products concentrations.

III. Results and disscutions

A. N_2 physisorption

MCF, MCF(N), MCF(B1) and MCF(B2) supports present type IV isotherms (Fig. 1). The observed hysteresis loops (IUPAC) are typical of mesoporous materials.

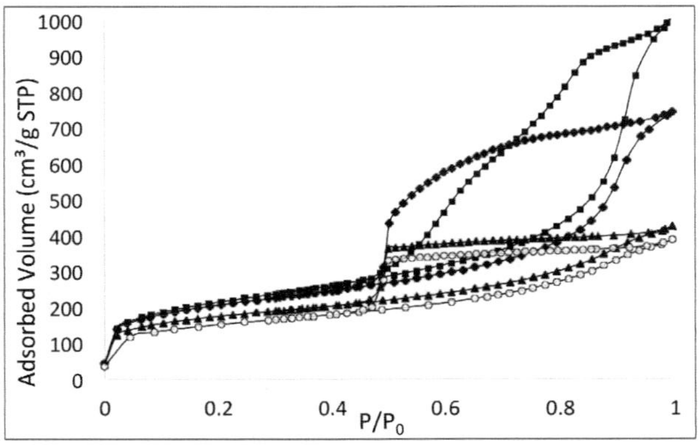

Fig. 1. N_2 adsorption/desorption isotherms at 77K of MCF(N) (■), MCF(♦), MCF(B1) (▲) and MCF(B2) (o).

Most of them are of type H2 or H1, which is in agreement with the expected presence, in mesocellular foams, of connected cells and of pores with different shapes.

Addition of ammonium fluoride or n-butanol during synthesis led to a modification of the pores volumes and of the dimensions of the cells/pores, but not as much as for the specific surface area values (Table.1). On one hand, the influence of NH_4F has been already described as being related to the mineralizing effect of fluoride ions, giving improved structural organization and higher pore volume [3]. On the other hand, n-butanol used as a cosurfactant in MCF(B1) and MCF(B2) led to quite low volumes values which turned to be even smaller than those obtained with SBA-15.

TABLE I. TEXTURAL PROPERTIES OF SOLIDS BEFORE AND AFTER Ni IMPREGNATION

Samples	Textural properties			
	S_{BET} (m^2/g)	V_{pores} (cm^3/g)	D_{cells} (nm)	$D_{Windows}$ (nm)
MCF	722	1.2	21	3.8
MCF(N)	755	1.6	21	5
MCF(B1)	696	0.7	15.2	4
MCF(B2)	687	0.6	7.0	4
Ni-MCF	594	0.9	18	3.8
Ni-MCF(N)	490	1.14	21	4.6
Ni-MCF(B1)	529	0.57	15.2	3.9

Due to the similar textural features of MCF(B1) and MCF(B2), further work in this paper will be focused only on MCF(B1) amongst these two samples. The shapes of the N_2 adsorption and desorption isotherms of the materials after impregnation of these supports with 5 wt.% Ni were similar to those before Ni addition, with type IV isotherms and H2 or H1 hysteresis loop (IUPAC) (Fig. 2). Nevertheless, the pores volumes and specific surface area decreased by 20 to 30% upon Ni insertion, the effect being the most pronounced in the case of Ni-MCF(N) (Table I). The diameter of the cells was retained in the case of Ni-MCF(N) and Ni-MCF(B1) samples but decreased a little bit, from 21 to 18 nm, for Ni-MCF. On the other hand, windows dimensions were reduced only for MCF(N). However, further characterization is needed in order to correlate the decrease of the dimensions of the cells or windows with the preferential location of the NiO particles.

B. X-Ray Diffraction (XRD)

Wide angle XRD patterns of the calcined Ni-supports are shown in Fig. 3. In all cases, a broad peak at about $2\theta = 22°$ is observed corresponding to diffusion phenomena in amorphous silica.

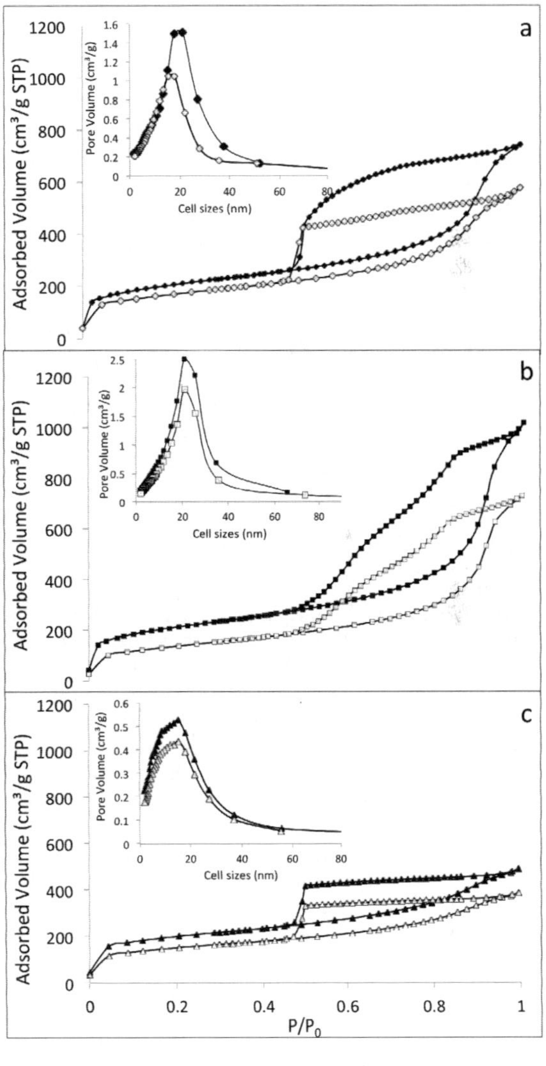

Fig. 2. N_2 adsorption/desorption isotherms at 77K and cell sizes distribution (from adsorption branch) for (a) : (♦) MCF, (◊) Ni-MCF, (b) : (■) MCF(N), (□) Ni-MCF(N) and (C) : (▲) MCF(B1), (Δ) Ni-MCF(B1).

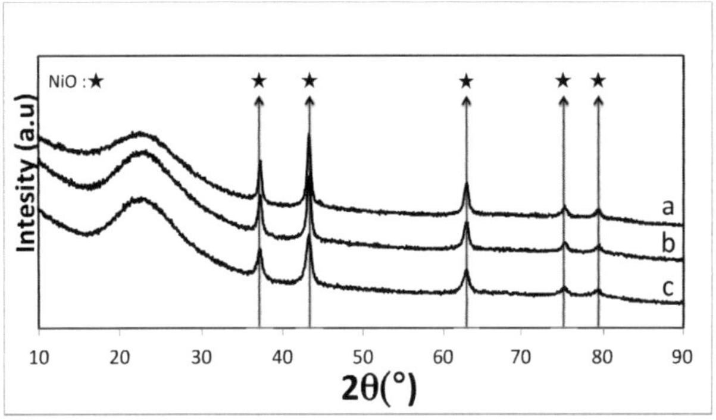

Fig. 3. XRD patterns of (a): Ni-MCF(N), (b): Ni-MCF, (c): Ni-MCF(B1) samples.

Five diffractograms peaks at 2θ = 37, 43, 62, 76 and 80° are also systematically seen and attributable to (111), (200), (220), (311) and (222) lattice planes of the face centered cubic structure of NiO oxide nanoparticles (JCPDS: 01-089-7130). The Scherrer formula was used to estimate the mean size of the NiO crystallites, assuming a spherical shape. The results are reported in Table II. With a 8.7 nm mean diameter, Ni-MCF(N) is characterized by the smaller crystallites.

This size is inferior to the critical maximum value (9 nm) reported by Kawi *et al* [9] as providing good performances in DRM reaction. Hence, the nanoparticles in Ni-MCF(N) could be used advantageously in the studied reaction. Conversely, it is commonly admitted that Ni metal particles having bigger size than the critical one are known to favor carbon deposition and hence enhance catalyst deactivation [9].

C. Morphological properties

Fig. 4. shows typical TEM images of Ni-MCF(N) which was characterized by the best textural properties both before and after Ni addition. The images are consistent with an homogeneous organization of the support and reveal rather homogenous dispersion of the NiO nanoparticles. With low magnification, rather big Ni deposits seem to be present (Fig. 4A) but they consist in aggregates of small Ni nanoparticles with mean diameters below 10 nm (Fig. 4B, B'). Very small isolated nanoparticles are also seen everywhere in the material, as is examplified in Fig. 4C (wight circles) and in the associated zoomed image (Fig. 4C') which also highlights the pores. The particle sizes estimated from TEM are thus in good agreement with those deduced from XRD data (Table II).

TABLE II. NiO CRYSTALLITES SIZE DETERMINED FROM XRD DATA ANALYSIS

Samples	**Ni-MCF**	***Ni-MCF(N)***	Ni-MCF(B1)
Mean NiO particle sizes (nm)	11.2	8.7	11.7

D. Reducibility

All calcined Ni-supports were reduced using H_2 under 650°C to convert NiO to Ni^0. TPR measurements were conducted in order to study the reducibility of deposited nickel in the calcined nickel mesocellular foams materials and indirectly its interaction with the support. Fig. 5 shows only one reduction peak per sample, centered at 405°C and 410°C for respectively Ni-MCF and Ni-MCF(B1), instead of ~ 470°C for Ni-MCF(N). Hence, both former samples behave similarly, whereas the occurrence of an H_2 consumption at higher temperature would suggest a stronger interaction between NiO and the support in the case of Ni-MCF(N). Nevertheless, peak surfaces are comparable for the three samples, in line with the similar Ni contents in all catalysts (5 wt %). Note also that the reduction is achieved for all samples before 600°C, confirming that the temperature of 650°C applied for *in situ* activation of the catalysts prior catalytic tests is appropriate for full metal reduction.

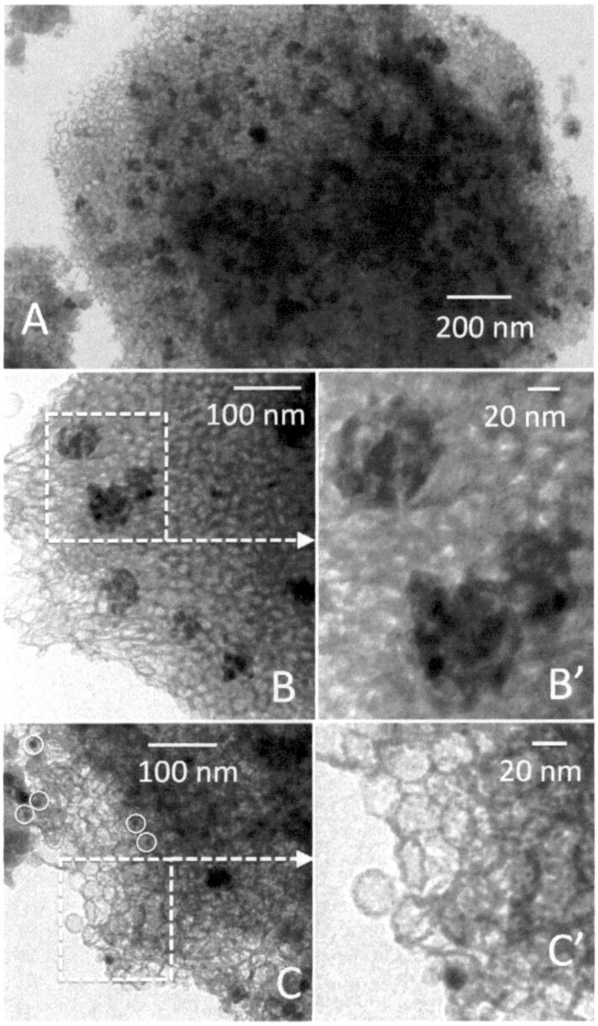

Fig. 4. TEM image of Ni-MCF(N) showing (A) a full grain and zones with (B) Ni aggregates or (C) isolated Ni nanoparticles. B' and C' are zoomed images of selected zones in B and C, respectively.

E. Catalytic performance in DRM

The Ni-MCF, Ni-MCF(N) and Ni-MCF(B1) samples were tested in DRM from 200 to 800°C. At low temperature, all catalysts are inactive, then the activity rises while increasing the temperature, in line with the endothermic nature of the reaction. Conversion of CH_4 and CO_2 vs. temperature are shown in Fig. 6, revealing significant differences in terms of catalytic performances between Ni-MCF, on one hand, and Ni-MCF(N) and Ni-MCF(B1), on the other hand.

By comparing the activity of the solids at 650°C, Ni-MCF showed the lower performances (42% CH_4 and 40% CO_2 conv.), while Ni-MCF(B1) (67% CH_4 and 75% CO_2 conv.) and Ni-MCF(N) (67% CH_4 and 68% CO_2 conv.) showed very good results. For the best catalysts, the conversions are very close to the expected thermodynamic values (maximum possible conversions) estimated with carbon formation. Not only the reactants conversions but also the selectivity of the catalysts in given products, especially their ability to give the desired H_2/CO ratio,is a very important criterion. Targeting the production of liquid fuels [12] for Fisher-Tropsch operations, a $H_2/CO=1$ is highly desirable. The values effectively obtained with the three selected materials are presented in Fig.7.

Hence, the less active Ni-MCF catalyst never gave the desired ratio, even at high temperature (770°C). In contrast, H_2/CO values very close to 1 were obtained with Ni-MCF(B1) at 650°C (0.98), 750°C (1.03) and 770°C (1.04) as well as with Ni-MCF(N) at 750°C (1) and 770°C (1.01). As a whole, the Ni-MCF(N) sample turned out to gather a lot of advantages such as the best textural properties, the smallest Ni crystallites size, the highest reduction temperature and good activity and selectivity.

For these reasons, this material was selected for a study of the stability at 650°C during 4 h in flowing reactant gases as above (Fig. 8).shows that Ni-MCF(N) is characterized by a really good stability vs. time at 650°C, working very closely to the thermodynamic conditions (with carbon formation), i.e. with conversion rates of CH_4 and CO_2 equal to 60 and 71%, respectively.

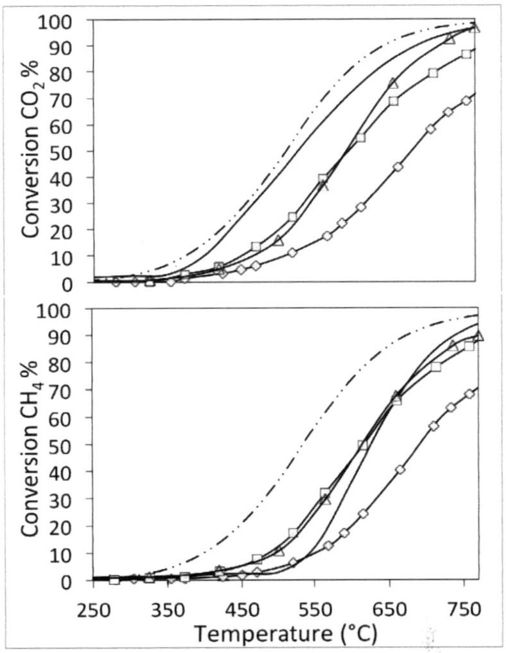

Fig. 6. Activity of reduced (◊): Ni-MCF, (□): Ni-MCF(N) and (Δ): Ni-MCF(B1) in DRM. Thermodynamic curves with (—) and without (-..) carbon formation are also mentioned.

In addition, the XRD pattern (Fig. 9.) of Ni^0-MCF(N) after use (spent catalyst) (a) is very similar to that after reduction (b), with three peaks at $2\theta = 44°$, 52° and 76° (JCPDS: 01-070-1849) characteristic of the Ni^0 phase and width indicating crystallites size of ~ 8.3 nm in both cases, thus emphasizing the absence of Ni sintering. In addition, the diffractogram of the spent catalyst did not reveal any presence of detectable structured carbon deposit, suggesting, if formed, that carbon phases remain limited.

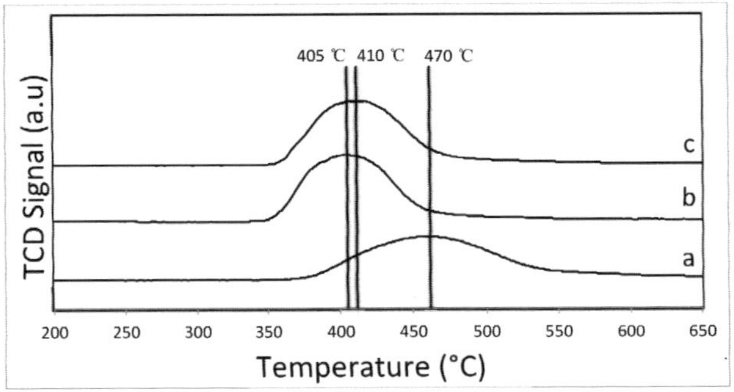

Fig. 5. Temperature programed reduction of: (a): Ni-MCF(N), (b): Ni-MCF, (c): Ni-MCF(B1) samples.

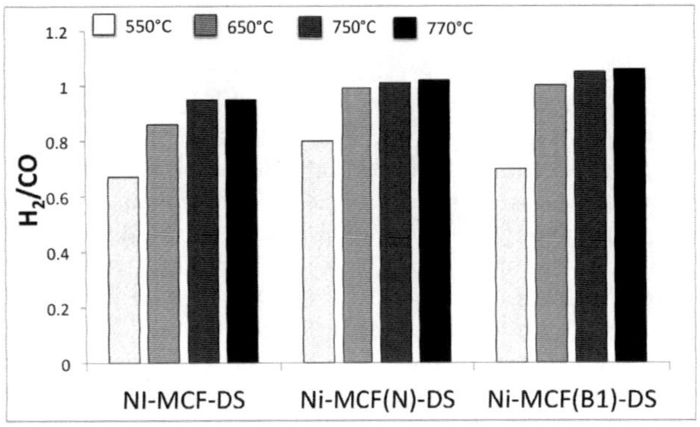

Fig. 7. H_2/CO ratio vs temperature for Ni-MCF, Ni-MCF(N) and Ni-MCF(B1) samples .

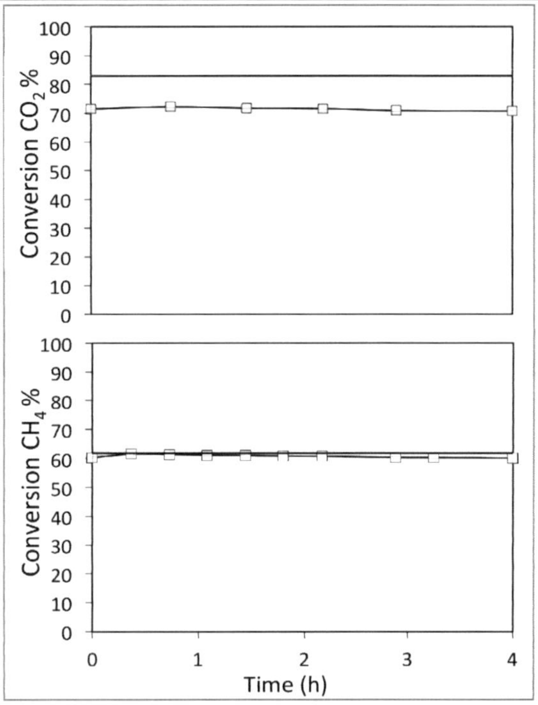

Fig. 8. Stability of CO_2 and CH_4 conversion in the presence of Ni-MCF(N). ( ). Thermodynamic curves with carbon formation (—) are also mentioned.

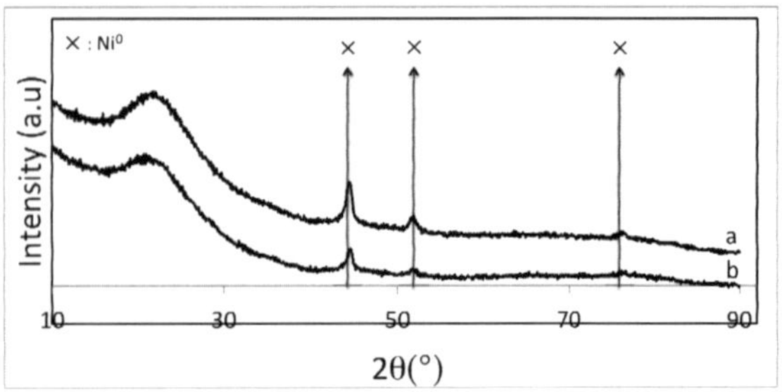

Fig. 9. XRD patterns for (a) spent Ni^0-MCF(N) and (b) Ni^0-MCF(N) after reduction.

IV. Conclusion

Silica mesocellular foams (MCF) with different textural properties were synthesized and used as nickel supports for preparing dry methane reforming catalysts. Unlike what was expected, silica samples synthesized using butanol as a co-surfactant (MCF(B1) and MCF(B2) materials) did not display the optimal textural properties. The largest windows apertures, pores volumes and surface area were obtained with the MCF(N) support which synthesis involved NH_4F. After nickel addition using the double-solvents method (post-synthesis) then activation, Ni-MCF(N) also led to the best nickel support-interaction and best nickel dispersion. It also led to the best catalytic performances in terms of CO_2 and CH_4 conversions, close to maximum thermodynamic limitations and slightly above those obtained on the samples with MCF support synthesized in presence of n-butanol. Also, Ni-MCF(N) was slightly more selective towards the production of syngas with a H_2/CO ratio of 1 convenient for Fischer Tropsch. Compared to bibliography, those catalyst showed satisfying activity even slightly below that previously obtained on a Ni-SiO_2 catalyst with the same Ni content [12]. Characterization data obtained on the spent catalyst agreed with very little carbon formation during the tests and the absence of metal active phase sintering.

Acknowledgment

The authors sincerely acknowledge the ERANETMED EU-FP7 initiative and the national ANR (France) and CNRS-L (Lebanon) agencies for funding of SOL-CARE (Energy-065) project (JC-ENERGY-2014 first call). Also, the authors thanks Sandra Casale (UPMC) for MET images and Hala Fallah (UOB) for XRD recorging.

References

[1] W. Shan, W. Wang, and H. Ru, "Siliceous mesocellular foams modified via a partitioned cooperative self-assembly process using hexane as pore swelling agent." J. Non-Cryst. Solids, vol. 425, pp. 183-189, 2015.

[2] J. S. Lettow, Y. J. Han, P. Schmidt-Winkel, P. Yang, D. Zhao, G. D. Stucky and J. Y. Ying, "Hexagonal to Mesocellular Foam Phase Transition in Polymer-Templated Mesoporous Silicas," Langmuir, vol. 16, pp. 8291-8295, 2000.

[3] P. Schmidt-Winkel, C. J. Glinkas and G. D. Stucky, "Microemulsion Templates for Mesoporous Silica," Langmuir, vol. 16, pp. 356-361, 2000.

[4] K. Szymanska, J. Bryjak, J. MrowiecbBialon, and A.B. Jarzebski, "Application and properties of siliceous mesostructured cellular foams as enzymes carriers to obtain efficient biocatalysts," Micropor. Mesopor. Mat, vol. 99, pp. 167-175, 2007.

[5] S. G. de Ávila, L. C. C. Silva and J. R. Matos, "Optimisation of SBA-15 properties using Soxhlet solvent extraction for template removal," Micropor. Mesopor. Mat, vol. 234, pp. 277-286, 2016.

[6] P. Y. Feng, X. Bu, G. D. Stucky, and D. J. Pine, "Monolithic Mesoporous Silica Templated by Microemulsion Liquid Crystals," J. Am. Chem. Soc, vol. 122, pp. 994-995, 2000.

[7] X. Zhou, A. Duan, Z. Zhao, Y. Gong, H. Wu, J. Li, Y. Wei, G. Jiang, J. Liu and Y. Zhang, "Synthesis of hierarchically porous silicas with mesophase transformations in a four-component microemulsion-type system and the catalytic performance for dibenzothiophene hydrodesulfurization," J. Mater. Chem. A 2, vol. 19, pp. 6823-6833, 2014.

[8] D. Pakhare and J. Spivey, " A review of dry (CO_2) reforming of methane over noble metal catalysts," Chem. Soc. Rev, vol. 43, pp. 7813-7837, 2014.

[9] S. Kawi, Y. Kathiraser, J. Ni, U. Oemar, Z. Li, and E. T. Saw, " Progress in Synthesis of Highly Active and Stable Nickel- Based Catalysts for Carbon Dioxide Reforming of Methane," ChemSusChem, vol. 8, pp. 3556-3575, 2015.

[10] J. van der Meer, I. Bardez-Giboire, C. Mercier, B. Revel, A. Davidson, and R. Denoyel," Mechanism of Metal Oxide Nanoparticle Loading in SBA-15 by the Double Solvent Technique," J. Phys. Chem. C, vol. 114, pp. 3507-3515, 2010.

[11] R. N. Widyaningrum, T. L. Church, M. Zhao, and A. T. Harris, "Mesocellular-foam-silica-supported Ni catalyst: Effect of pore size on H_2 production from cellulose pyrolysis," Int. J. Hydrogen. Energ, vol. 37, pp. 9590-9601, 2012.

[12] M. N. Kaydouh, N. El Hassan, A. Davidson, S. Casale, and P. Massiani, "Highly active and stable N/SBA-15 catalysts prepared by a ''two solvents'' method for dry reforming of methane," Microp. Mesop. Mater, vol. 220, pp. 99-109, 2016.

Ultrasound and ionic liquid-enhanced extractive desulfurization of diesel

Karma Zuraiqi, Paul Nancarrow,* Hussain Ahmed
Department of Chemical Engineering
American University of Sharjah
United Arab Emirates
pnancarrow@aus.edu

***Abstract*—The current desulfurization technology, hydrodesulfurization (HDS), is energy intensive, uses large quantities of hydrogen, and is ineffective at removing some sulfur compounds. Ionic liquids have been recently shown to perform well as extraction solvents for the removal of sulfur from diesel. However, a major limitation towards their industrial use is the low mass transfer rates that result from their relatively high viscosities. This study investigates the use of ultrasound to overcome this problem. The ionic liquid, 1-ethyl-3-methylimidazolium acetate ([C_2mim][CH_3COO]), was used to desulfurize model diesel consisting of dibenzothiophene in hexadecane, in the absence and presence of ultrasound. It was found that sonification can greatly enhance the rate of mass transfer of DBT from diesel to the IL, resulting in a 57% sulfur removal for a 75% decrease in extraction time for batch studies, which results in a reduction in the number of equilibrium stages necessary for a continuous extraction process.**

Keywords—ionic liquid; ultrasound; desulfurization; diesel

I. INTRODUCTION

Today diesel fuel plays a crucial role in shaping our economies worldwide. It is used in transportation, industrialization and power generation, for example [1]. Diesel consumption is directly related to global economy, making its optimization a universally central issue. However, the widespread use of diesel has also associated with atmospheric pollution, environmental degradation and serious health issues. In recent years, the need for improved pollution control systems is becoming more popular due to increasing environmental awareness and increase in stringent regulations regarding fuel oils quality and specifications. Sulfur compounds present in diesel turn to SOX derivatives upon combustion, which have major detrimental environmental impacts. These compounds cause acid rain, ozone depletion, and jeopardize living organisms. Acid rain can lower soil pH, affect agriculture and cause deforestation, change climate, endanger marine life, and corrode or weaken materials. Sulfur dioxide, if inhaled, can inflict serious health problems. Health problems associated with sulfur dioxide generally involve the respiratory system. Current regulations call for ultra-clean fuel, with sulfur content as low as 10 ppm in many countries [2].

The current industrial desulfurization technology, HDS is a costly, capital and energy intensive process requiring large quantities of explosive hydrogen, and is not highly effective at removing sterically hindered organosulfur compounds such as DBT and its derivatives, which are common in raw diesel. The problems with the current technology have brought about the need for innovative cost effective and safer alternative solutions.

Owing to the negligible vapor pressures of ILs and their effectiveness in the removal of DBT, ILs make for a very attractive solvent choice in the extractive desulfurization of diesel fuel [3]. However, the high viscosity of these liquids, which results in reduced mass transfer rates can be considered a significant disadvantage. Hence, desulfurization can be aided by auxiliary techniques that could enhance mass transfer.

The ability of US to agitate, generate cavitation, cause localized increases in temperatures, and attenuate solvent viscosity [4] are the reasons behind the selection of irradiation in this study to facilitate the desulfurization process and overcome the mass transfer limitations in IL extraction. The effects of US on the mass transfer rate and extraction process design in the desulfurization of diesel are thus investigated in this paper.

II. EXPERIMENTAL

A. Chemicals and Materials

Dibenzothiophene and n-hexadecane, were supplied by the Sigma-Aldrich Chemical Company with cited purities of 98%, and 99% respectively. [C_2mim][CH_3COO] was supplied by IOLITEC Ionic Liquids Technologies with cited purity of 95%. Supplied chemicals were used without any further purification.

B. Desulfurization Experiment

The initial sulfur content of model diesel was recorded before equal volumes of model diesel and IL were placed into the mixing vessel. The system was subjected to sonification in a Bransonic 3510 40 Hz ultrasonic batch at room temperature for a fixed time duration. Samples of the diesel phase were extracted at fixed time intervals and sulfur content was determined using an Ametek X-ray fluorescence (XRF) spectrometer. To ensure complete phase separation prior to analysis, samples were subjected to centrifugation for several minutes and then immediately tested using XRF. An identical system without sonification was set up as a control to compare the rate of desulfurization with, and in the absence of ultrasound, as illustrated in Fig. 1.

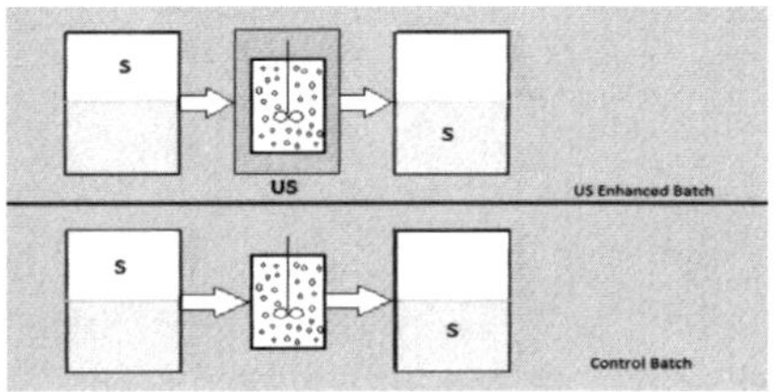

Fig. 1 Experimental setup for desulfurization with and without sonification.

III. Results and Discussion

A. Effect of Ultrasound on Desulfurization

By investigating mass transfer rate in two batch desulfurization systems, one batch subjected to ultrasound irradiation while maintaining the other batch as a control, the influence of ultrasound on the desulfurization rate was studied. Both systems were operated under the same conditions: room temperature at 22 °C, model diesel to IL ratio of 1:1, same vessel size. In the first experiment, no agitation was used, so that the interfacial boundary was clearly defined.

As evident from Fig. 2, the rate of desulfurization in the system involving sonification was notably higher than that without sonification, indicating the potential for using ultrasound to enhance the extraction process.

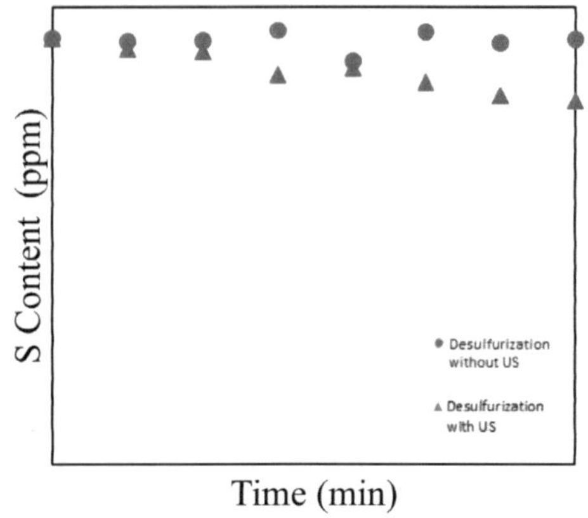

Fig. 2. Change in sulfur content of model diesel as a function of time for non-agitated batch extraction with [C_2mim][CH_3COO] in the presence of sonification (blue) and without sonification (red) in a 31 mm diameter vessel.

Results illustrated in Fig. 2 indicate that while the system is undergoing a steady desulfurization over time, changes in sulfur content for the system not subjected to ultrasound are almost negligible. This illustrates a very low rate of mass transfer in the IL phase and, hence, the need for a mass transfer enhancing method. In contrast, in the sonicated system, sulfur content is clearly decreasing over time.

Fig. 3 illustrates the significant improvement in desulfurization achieved through sonification in a larger extraction vessel (67 mm diameter). Both systems show faster sulfur removal over time compared with the 31mm diameter extraction vessel; this is attributed to to the larger interfacial area. Furthermore, the sonicated process has removed approximately three times more sulfur compared with the non-sonicated process after 180 minutes.

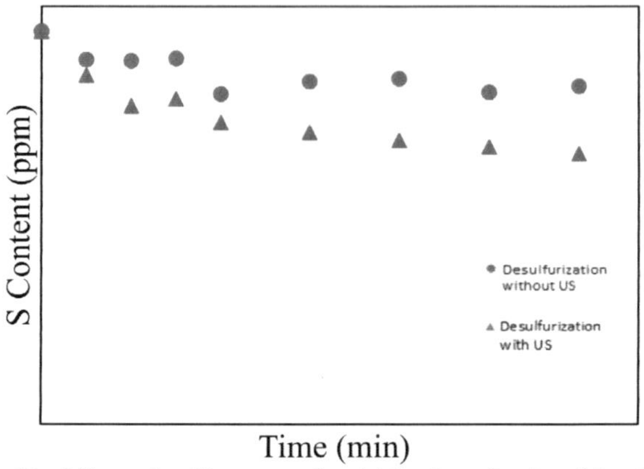

Fig. 3 Change in sulfur content of model diesel as a function of time for a non-agitated batch extraction using [C_2mim][CH_3COO] in the presence of sonification (blue - Δ) and without sonification (red - o) in a 67 mm diameter vessel.

B. Effect of Agitation

As shown in the previous section, the interfacial area between the liquid phases is a key parameter affecting mass transfer rate. A common method to increase interfacial area in extraction systems is mechanical agitation. Therefore, this section addresses the effect of agitation on the rate of desulfurization of diesel.

To investigate the use of mechanical agitation on the desulfurization rate, in the presence and absence of ultrasound, experiment was repeated under identical conditions except that an impeller was used to provide mixing at 400 rpm (Fig. 4).

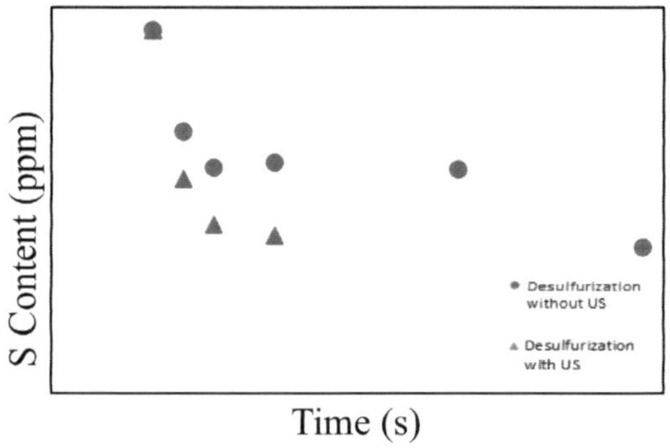

Fig. 4 Change in sulfur content of model diesel as a function of time for the batch extraction with [C_2mim][CH_3COO] with agitation at 400 rpm in the presence of sonification (blue - Δ) and without sonification (red - o) in a 67 mm diameter vessel.

With the introduction of agitation, the system reached equilibrium at a much faster rate, i.e. reaching an equilibrium sulfur removal of ~60% in seconds, rather than hours. However, as shown in Fig. 4, the use of ultrasound can significantly enhance the mass transfer rate even when mechanical agitation is used. Consequently, these results verify that ultrasound can be potentially integrated into extractive desulfurization involving IL as extractive solvents, to significantly accelerate desulfurization rates.

The benefit of agitation is further illustrated by Fig. 5, which compares the agitated with non-agitated system for the US assisted desulfurization of model diesel using IL. As expected, desulfurization is greatly enhanced by the agitation to the system.

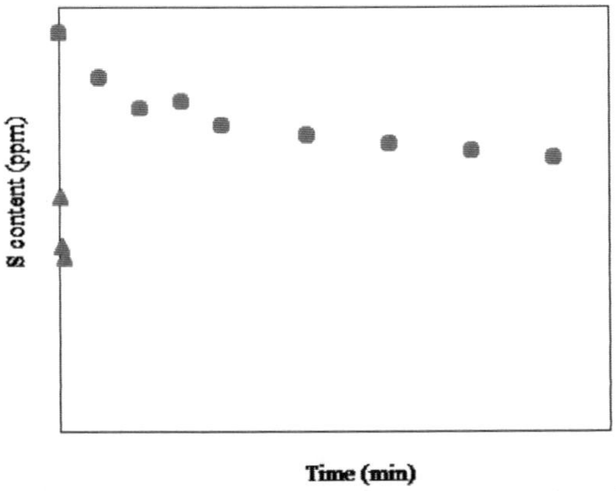

Fig. 5 Effect of agitation on desulfurization for batch extraction subjected to sonification, for an agitated (blue - Δ) and a non-agitated (red - o) system, in a 67 mm diameter vessel.

C. *Effect of Vessel Size*

Vessel size is another system parameter that can affect the interfacial area between the two liquid phases. Fig. 6 illustrates the effect of vessel diameter on the extractive desulfurization of diesel using [C_2mim][CH_3COO].

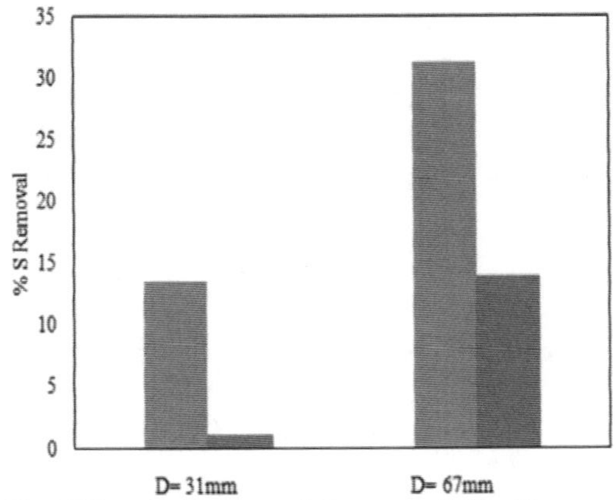

Fig. 6 Effect of extractor vessel diameter on the % sulfur removal for non-agitated batch extraction, in the presence of sonification (blue) and without sonification (red) in 31 mm and 67 mm diameter vessels for 180 min extraction time.

For a fixed desulfurization period, the employment of a larger vessel diameter shows an increase in sulfur removal of approximately 18% and 13% in the presence and absence of sonification respectively. For the non-agitated system, interfacial area is simply the cross-sectional area of the vessel, which increases with vessel diameter. Therefore, the results obtained are compliant with expectations.

IV. Extraction Process Design

A. *Effect of Ultrasound on Initial Mass Transfer Rate*

Since ultrasound has significant potential for the improvement of mass transfer in IL extraction processes, quantification of the mass transfer rate is necessary for the adequate design and scale-up of process equipment.

The initial mass transfer rate can be acquired from the slope of the tangent line formed for the region of the curve at the early stages of desulfurization, as illustrated in Fig. 7.

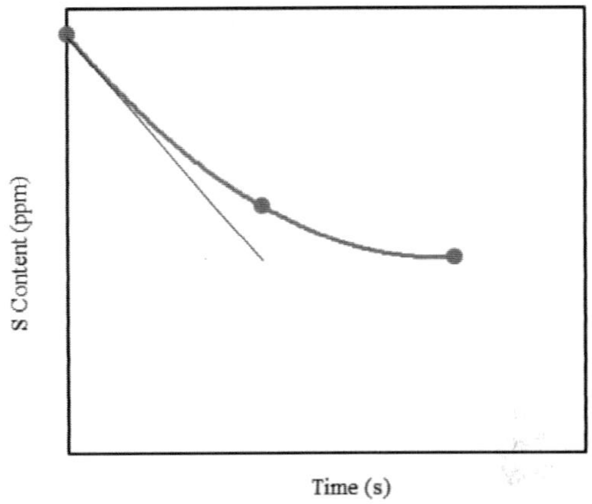

Fig. 7 Schematic representation of the generation of initial mass transfer rate for the desulfurization experiment.

Calculated initial mass transfer rates for the desulfurization of agitated batch experiments mixed at 400 rpm at room temperature are compared to show the effect of ultrasound. As evident from Table I, the initial mass transfer rate and volumetric mass transfer coefficient, KA, are almost doubled through the use of ultrasound, as excepted.

TABLE I. Initial mass transfer rate for the extraction of DBT from hexadecane in the presence and absence of ultrasound

	Ultrasound Assisted	*No Ultrasound*
Initial Mass Transfer Rate (ppm/s)	117	65
$KA\text{x}10^5 (1/s)$	1.78	0.99

Although volumetric mass transfer coefficients can be derived from the initial mass transfer rates, the use of an interphase mass transfer model that can adequately explain the mass transfer in batch extraction can provide more accurate representation.

B. *Mass Transfer Model*

In order to quantify the effect of US on the desulfurization process, and to calculate the mass transfer coefficient for sonicated systems as well as systems not subjected to US, mass transfer rates were modeled. Equation (1) describes the mass transfer in the batch extraction system:

$$-V_{C16}\frac{dC_{C16}}{dt} = KA_i\left(C_{C16} - \left(\frac{V_{C16}}{D_rV_{IL}}\right)\left(C^0_{C16} - C_{C16}\right)\right) \tag{1}$$

where C_{C16} is the sulfur content in the hexadecane, C^0_{C16} is the initial sulfur content at the initial time, V_{C16} is the hexadecane volume, V_{IL} is the IL volume, A_i is the interface area, D_r is the distribution coefficient, and K is the mass transfer coefficient of the system. Since the volume of the IL is equal to the volume of the hexadecane, i.e. $V_{C16} = V_{IL}$, the equation can be simplified.

By separation of the variables, the solution to the differential equation was obtained which was then fitted to the experimental data using the nlinfit toolbox in MATLAB. The toolbox utilizes the free variable available in the power of the exponential to fit the data for the system.

$$C_{C16} = C^0_{C16}\left(\left(\alpha - \frac{1}{\alpha}\right) + \frac{e^{-\alpha\beta t}}{\alpha}\right) \tag{2}$$

where,

$$\alpha = 1 + \frac{1}{D_r} \tag{3}$$

and

$$\beta = \frac{KA_i}{V_{C16}} \tag{4}$$

From analysis of the fitted parameters, the volumetric mass transfer coefficient, KA, was obtained for each system. This is a quantitative measure of the desulfurization rate in the extraction process and can be used to compare the various systems. This model was plotted on the same axes as the experimental data in Figs 8, 9, and 10, to show the performance of the model in representing the mass transfer behavior of the system.

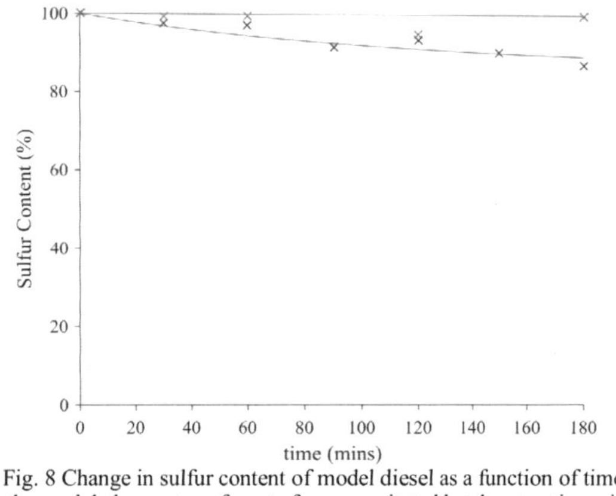

Fig. 8 Change in sulfur content of model diesel as a function of time using the modeled mass transfer rate for non-agitated batch extraction with [C_2mim][CH_3COO] in the presence of sonification (blue) and without sonification (red) in a 31 mm diameter vessel.

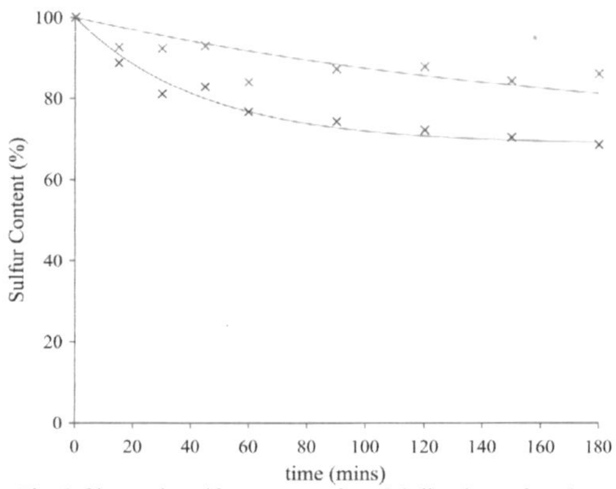

Fig. 9 Change in sulfur content of model diesel as a function of time using the modeled mass transfer rate for non-agitated batch extraction with [C_2mim][CH_3COO] in the presence of sonification (blue) and without sonification (red) in a 67 mm diameter vessel.

Volumetric mass transfer coefficients derived from Figs. 6 and 7 are summarized in Table II for non-agitated batch extraction.

TABLE II. EFFECT OF ULTRASOUND ON VOLUMETRIC MASS TRANSFER COEFFICIENT FOR TWO VESSEL SIZES

	67mm (KAx10^5(1/s))	*31mm (KAx10^5(1/s))*
Ultrasound Assisted	38.1	6.77
No Ultrasound	8.71	0.26

Table II shows that the volumetric mass transfer coefficient for the non-agitated desulfurization system increases with the increase in the extractor vessel diameter, as well as the use of ultrasound.

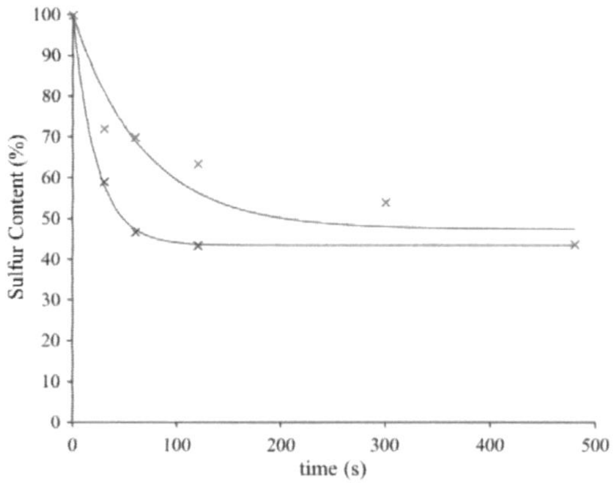

Fig. 10 Change in sulfur content of model diesel as a function of time using the modeled mass transfer rate for batch extraction agitated at 400 rpm with [C2mim][CH3COO] in the presence of sonification (blue) and without sonification (red) in a 67 mm diameter vessel

The enhancement in rate of sulfur removal due to sonification can be clearly observed in Fig. 10, showing that the system subjected to ultrasound reaches the equilibrium sulfur concentration faster than the system without ultrasound. Moreover, using the model illustrated in Fig. 8 the volumetric mass transfer coefficients are calculated for comparison between the irradiated system and the system without ultrasound.

TABLE III. EFFECT OF ULTRASOUND ON VOLUMETRIC MASS TRANSFER COEFFICIENT FOR AN AGITATED BATCH EXTRACTION

	$KA \times 10^2 (1/s)$
Analytic	2.87
No Ultrasound	2.80
Ultrasound Assisted	8.43

Analytic solution values presented in Table III are used as a reference to which calculated volumetric mass transfer coefficient values for the agitated system are compared. The analytic values are determined by calculating the diffusivity using the Wilke-Chang correlation, D_{WC}, and diffusivity using the Schiebel correlation, D_{Sch}, as summarized in equations (5-11) [5].

$$D_{WC} = \frac{7.4 \times 10^{-8} (\emptyset Mwt_{IL})^{0.5} T}{\mu_{IL} V_{DBT}^{0.6}} \quad (5)$$

$$D_{Sch} = 8.2 \times 10^{-8} \left(1 + \left(\frac{3V_{C_{16}}}{V_{DBT}}\right)^{0.66}\right) \frac{T}{\mu_{C_{16}} V_{DBT}^{0.33}} \quad (6)$$

$$Re = \frac{N d_i \rho}{\mu} \quad (7)$$

$$Sc = \frac{\mu}{\rho D_{WC/Sch}} \quad (8)$$

$$Sh = 0.664 Re^{0.5} Sc^{\frac{1}{3}} \quad (9)$$

$$Sh = \frac{k_{IL/C_{16}} d_i}{D_{WC/Sch}} \quad (10)$$

$$\frac{1}{K} = \frac{1}{k_{C_{16}}} + \frac{1}{d_r k_{IL}} \quad (11)$$

Where Sc, Sh, and Re are the Schmidt, Sherwood, and Reynolds numbers respectively. Other parameters include Mwt_{IL}, molecular weight of the IL, T, system temperature, μ, fluid viscosity, ρ, fluid density, d_i, impeller diameter, d_r, distribution coefficient, and N which represent the molecular weight of the IL, system temperature, fluid viscosity, fluid density, impeller diameter, distribution coefficient, and impeller speed respectively. Here for the IL/oil system, the association factor, ϕ, is assumed to equal to 1.0 due to IL characteristics. It can be observed from equations (5-11) that K values can be affected by system diffusivities and solution properties.

Mass transfer as per Tables II and III has been consistently enhanced in the presence of sonification, with results shown in Table III showing a KA value for the ultrasound assisted system that is more than 3 times the KA value for the system not subjected to ultrasound. Moreover, vessel size and agitation further improved the mass transfer coefficient. Therefore, the increase in the volumetric mass transfer coefficient, and the enlargement of the interfacial area due to sonification result in the acceleration of desulfurization rates and total sulfur removal. It is well established that ultrasound can enhance extraction rates due to either increases in mass transfer coefficient and effective diffusivity [6], or by decreasing dispersed droplets size and increasing interfacial area [7]. Observed overall enhancement to the desulfurization of diesel using $[C_2mim][CH_3COO]$, brought about by sonification, can be attributed to a combination of both factors.

V. CONCLUSION

As extraction solvents, ILs show great extraction potential in desulfurizing diesel oil. However, the environmental compatibility of these liquids is faced by operation difficulty due to their high viscosity. Lower extraction rates due to the high viscosity characteristic, nonetheless, can be resolved by applying techniques of mass transfer enhancement. This paper proposes the use of ultrasound to accelerate desulfurization rates.

Results obtained indicate significant mass transfer improvement by sonification, showing a 57% sulfur removal for a 75% reduction in residence time for the agitated batch extraction upon subjection to sonification. Furthermore, volumetric mass transfer coefficients obtained for the ultrasound assisted extraction are consistently greatly larger than those for the non-ultrasound system. The increase in volumetric mass transfer coefficient and interfacial areas displayed by the system subjected to ultrasound result in an enhancement to the volumetric mass transfer rate.

References

[1] US Energy Information Administration. https://www.eia.gov/energyexplained/index.cfm?page=diesel_use

[2] L.G. Chestnut, D. M. Mills, "A fresh look at the benefits and costs of the US acid rain program," *J. Environ. Manage.*, vol. 77, pp. 252-266, 2005.

[3] P. Nancarrow, N. Mustafa, A. Shahid, V. Varughese, U. Zaffar, R. Ahmed, N. Akther, H. Ahmed, I. AlZubaidy, S. Hasan, Y. Elsayed, Z. Sara, "Technical evaluation of ionic liquid-extractive processing of ultra low sulfur diesel fuel," *Ind. Eng. Chem. Res.*, vol. 54, pp. 10843-10853, 2015.

[4] B. Zisu, M. Schleyer, and J. Chandrapala, "Application of ultrasound to reduce viscosity and control the rate of age thickening of concentrated skim milk," *Int. Dairy J.*, vol. 31, no. 1, pp. 41–43, 2013.

[5] A. R. Ferreira *et al.*, "Removal of thiols from model jet-fuel streams assisted by ionic liquid membrane extraction," *Chemical Engineering Journal*, vol. 256, pp. 144-154, 2014.

[6] L. H. Thompson and L. K. Doraiswamy, "Sonochemistry: Science and Engineering," Industrial & Engineering Chemistry Research, vol. 38, no. 4, pp. 1215-1249, 1999.

[7] J. Huang, Y. Liu, Z. Song, Q. Jin, and X. Wang, "Kinetic study on the effect of ultrasound on lipase-catalyzed hydrolysis of soy oil: Study of the interfacial area and the initial rates," *Ultrasonics sonochemistry*, vol. 17, no. 3, pp. 521-5, 2010.

Evaluation of Polymer Alternating Waterflooding in Multilayered Heterogeneous Waterflooded Reservoir

Warut Tuncharoen
Department of Mining and Petroleum Engineering
Chulalongkorn University
Bangkok, Thailand
Email: warut.t@student.chula.ac.th

Falan Srisuriyachai
Department of Mining and Petroleum Engineering
Chulalongkorn University
Bangkok, Thailand
Email: falan.s@chula.ac.th

***Abstract*—Polymer flooding is widely implemented to improve oil recovery since polymer can increase sweep efficiency and smoothen heterogeneous reservoir profile. However, polymer solution is somewhat difficult to be injected due to high viscosity and thus, water slug is recommended to be injected before and during polymer injection in order to increase an ease of injecting this viscous fluid into the wellbore. In this study, numerical simulation is performed to determine the most appropriate operating parameters to maximize oil recovery. The results show that pre-flushed water should be injected until water breakthrough while alternating water slug size should be as low as 5% of polymer slug size. Concentration for each polymer slugs should be kept constant and recommended number of alternative cycles is 2. Combining these operating parameters altogether contributes to oil recovery of 53.69% whereas single-slug polymer flooding provides only 53.04% which is equivalent to 8,000 STB of oil gain.**

Keywords—Polymer flooding; Polymer Injectivity; Concentration Sorting; Number of Alternative Cycle

I. INTRODUCTION

In recent years, oil price has plunged below satisfaction for the oil producers which triggered them to save any cost as much as possible especially, the cost of exploring and drilling new wells so, various oil recovery techniques are developed to extend the oil production life. Waterflooding used to be a common technique to increase the amount of oil owing to its simplicity and cost-effectiveness. However, waterflooding has one major problem which happens when water bypasses oil to the highest permeability channel, leaving some amounts of oil non-displaced due to unfavorable mobility of water compared to that of oil.

Instead of solely water, one benefit of polymer over water is that it creates more viscous fluid. Thus, when polymer solution is injected into the reservoir, polymer solution with higher viscous force will sweep more oil to the production well comparing to conventional waterflooding. Another benefit of polymer solution is that certain types of polymer tend to be adsorbed onto rock surface which results in reduction of the flow area of water, causing relative permeability to water to reduce and consequently, sweep efficiency is increased and more oil is produced.

However, the injection of solely polymer solution into a well still has a drawback. A major problem is that the ability to inject polymer into the reservoir as known as "injectivity" is low since polymer itself is very viscous so, the desired injection rate is scarcely achieved. Moreover, high adsorption of polymer onto rock surface could cause adverse effect on high reduction of flow ability. Therefore, water should be injected in between to increase the injectivity of polymer. Moreover, water would result in desorption of polymer on the rock surface which provides more ease for injecting the following polymer slugs.

II. POLYMER FLOODING IN HETEROGENEOUS RESERVOIR

A. Mobility Control Mechanism

Typically, polymer has two major characteristics affecting the increment of oil production; first, polymer is more viscous than water, when it is mixed with water, viscosity of water is increased. This reduces mobility ratio between water and oil to favorable condition which is when mobility ratio is less than one. Second, polymer tends to adsorb onto rock surface, thus, water has difficulty traveling through pore spaces, causing relative permeability to water to decline and turning mobility ratio to a favorable condition also.

Mobility ratio can be obtained from the following equation:

$$M = \frac{\lambda_w}{\lambda_o} = \frac{k_{rw}/\mu_w}{k_{ro}/\mu_o} \quad (1)$$

When, M = Mobility Ratio

λ_w = Mobility of Water

λ_o = Mobility of Oil

k_{rw}= Relative Permeability to Water

k_{ro}= Relative Permeability to Oil

μ_w= Viscosity of Water

μ_o= Viscosity of Oil

Mobility ratio implies relative speed between two fluids. In case of unfavorable condition (M>1), water travels faster than oil and tends to breakthrough first, on the other hand, oil travels faster than water in case of favorable condition (M<1).

B. Relative Permeability Reduction

Apart from increasing the viscosity of water, polymer solution also has effect on reduction of relative permeability to water without affecting the relative permeability to oil [1]. Considering mobility ratio equation, when relative permeability to water decreases, mobility ratio tends to decrease and converge to a more favorable condition (M<1).

There are two main measures in permeable media flow which are resistance factor (RF) and residual resistance factor (RRF)

1. Resistance Factor (RF): a relative pressure drop caused by polymer flooding and waterflooding. By using Darcy's Law, this term can be written as a mobility ratio between water and polymer as follows:

$$RF = \frac{\Delta p_{polymer}}{\Delta p_{water}} = \frac{\lambda_{water}}{\lambda_{polymer}} \tag{2}$$

2. Residual Resistance Factor (RRF): the ratio between relative permeability to water and water after polymer flooding, also known as "Permeability Reduction"

$$RRF = \frac{k_{rw,before.polymer.flooding}}{k_{rw,after.polymer.flooding}} \tag{3}$$

C. Polymer Flow Behavior in Porous Media

a. Inaccessible Pore Volume (IPV)

IPV happens when larger polymer molecule is introduced into the reservoir and cannot flow through smaller pore space of rock. The volume of uninvaded pore space is known as *Inaccessible Pore Volume (IPV)* [2]. Factors that affect *IPV* include polymer molecular weight, permeability-porosity of rock and pore size distribution.

b. Polymer Retention

Polymer retention consists of three main mechanisms which are polymer adsorption, mechanical entrapment and hydrodynamic retention.

1. Polymer Adsorption: the charge interaction between polymer and rock surface causes polymer molecule to be adsorbed onto rock surface. Thus, some part of polymer mass will be removed from bulk polymer solution in this interaction. Polymer adsorption depends upon surface area of rock exposed to polymer solution.

2. Mechanical Entrapment: this mechanism contributes a major part of polymer retention. It usually occurs in low-permeability formation. Since polymer is flexible, when it is trapped in a pore channel, a fluid can still flow through a polymer chain and brings even more polymer into a channel, resulting in an accumulation of polymer in a channel.

3. Hydrodynamic Retention: this mechanism is similar to mechanical entrapment except that trapped polymer molecule can be released when fluid does not flow through pore channel anymore [3].

D. Reservoir Heterogeneity

Reservoir Heterogeneity is defined as the variation of rock properties in different locations within the reservoir. One of the significant rock properties affecting heterogeneity is permeability. Combination of permeability can be classified into Parallel Layer and Serial Layer. Moreover, the heterogeneity of reservoir can be quantified by several methods. One of the widely used method is Lorenz Coefficent (L_k) which can be acquired from Fig. 1 by ratio of area 1 and area 2. If L_k is close to zero, reservoir is considered to have a uniform permeability distribution [4].

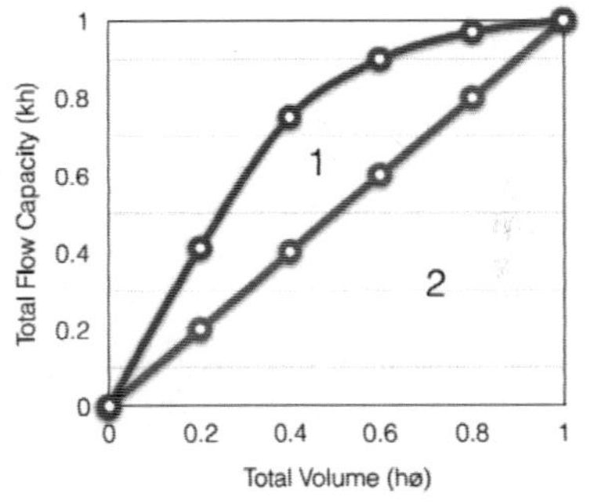

Fig. 1. Flow Capacity Distribution for determination of Lorenz Coefficient

III. METHODOLOGY

After theory of polymer flooding has been reviewed, all the procedures to achieve the purpose of this study are listed and described in this section.

1. Construct heterogeneous model with single Lorenz coefficient at L_k=0.275 with coarsening upward sequence

2. Perform waterflooding from the start of oil production to be a reference for other polymer flooding cases

3. Perform single-slug polymer flooding with polymer concentration of 1,000 ppm and 0.25 PV slug size to determine:

- *Polymer Injection Starting Time:* first day of production, after injecting 25% of water before breakthrough, after injecting 50% of water before breakthrough, at water breakthrough, after watercut reaches 25%, and after watercut reaches 50%.

4. Perform polymer alternating waterflooding. Keep the same injected polymer mass while varying the following operating parameters:

- *Concentration Sorting:* descending order, ascending order
- *Alternating Water Slug Size:* 5%, 10%, 25%, 50%, 100% (same size) of polymer slug size
- *Number of Alternative Cycles:* 1, 2, 3, 4, 5

5. Compare and analyze the results, using outcome parameters such as oil recovery factor, production time, etc.

Reservoir simulation model is generated using Cartesian grid model. The reservoir is heterogeneous with L_k of 0.275. Other important reservoir properties are shown in TABLE I as follows:

TABLE I. RESERVOIR PROPERTIES FOR BASE CASE MODEL

Parameters	***Values***	***Unit***
Grid dimension	33×33×9	Block
Grid size	20×20×12	ft.
Top of reservoir	7,000	ft.
Effective porosity (Ø)	20	%
Average permeability (k_v=k_h)	230	mD
Initial water saturation (S_{wi})	20	%
Reference pressure at datum depth	3,445	psia
Reservoir temperature	182	°F
Oil gravity	20	°API
Solution gas oil ratio (R_s)	50	scf/stb
Original oil in place (*OOIP*)	1,276.71	MMstb

Hole size for both injector and producer is 6 inches. For the injector, bottomhole pressure (BHP) is limited at 4,700 psi due to fracture pressure calculated using Hubbert-Willis equation at 4,743 psi and at producer; minimum bottomhole pressure is set at 420 psi beyond the bubble point pressure of 416 psi to continuously produce single-phase oil. Maximum injection and production rate are both operated at 800 STB/D to balance the reservoir pressure. Production will be terminated in case that watercut at producer reaches 90%. Other constraints are summarized in TABLE II.

TABLE II. PRODUCTION CONSTRAINTS FOR RESERVOIR SIMULATION

Parameters	***Values***	***Unit***
Fracture Pressure	4,743	psi
Maximum BHP at Injector	4,700	psi
Bubble Point Pressure	416	psi
Minimum BHP at Producer	420	psi
Maximum Injection Rate	800	STB/D
Maximum Production Rate	800	STB/D
Maximum Watercut at Producer	90	%

IV. RESULTS AND DISCUSSION

In order to identify appropriate operating parameters for polymer alternating waterflooding, including polymer injection starting time, alternating water slug size, concentration sorting and number of alternative cycles, first, waterflooding is performed as a base case for other polymer flooding cases. After that, several polymer flooding cases are simulated to determine the most appropriate scenario to perform polymer alternating waterflooding.

A. Waterflooding

Waterflooding is a conventional oil recovery method performed by injecting water into the reservoir to maintain the reservoir pressure. Waterflooding will be used as a reference to determine the benefit from injecting polymer into the reservoir.

Fig. 2 shows that water starts to breakthrough at 7th month of production which is very early because this reservoir is highly heterogeneous (L_k=0.275) and injection rate is quite high so, water tends to flow through the highest permeability channel only. In accordance with watercut, at the early time before breakthrough, pressure support from waterflooding can maintain oil rate at 800 STB/D later, after water breakthrough, oil rate drops rapidly due to insufficient pressure support from waterflooding, finally, oil production is terminated in mid-2005 due to watercut constraint with cumulative oil production of 0.49 MMstb and oil recovery factor of 38.8%.

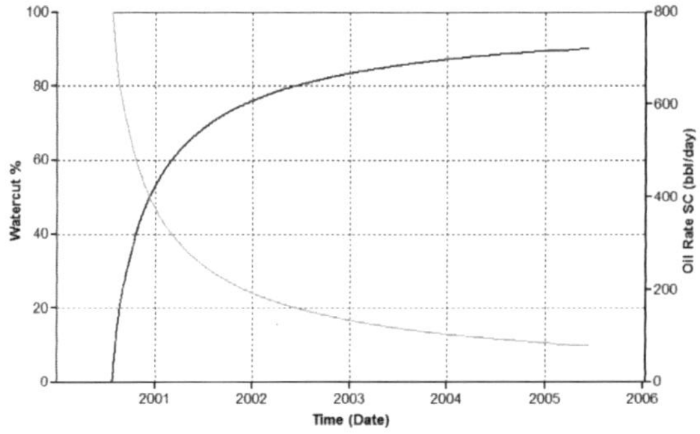

Fig. 2. Plot of watercut at producer (blue) on the left scale and oil rate (green) on the right scale as a function of time of waterflooding

B. Polymer Injection Starting Time

A quarter-five spot pattern is applied with the same production constraints used in waterflooding case. Hydrolized Polyacrylamide (*HPAM*) polymer is selected since it can exhibit both *water viscosity enhancement* and *relative permeability reduction* behaviors. *HPAM* has a molecular weight of 8,000 lb/mol. Single-slug polymer flooding is performed with polymer concentration of 1,000 ppm and 0.25 PV slug size. 20% *IPV* and *RRF* of 2 are assumed in this study.

One major problem of injecting solely polymer slug is that liquid injection rate cannot be attained at desired rate (800 STB/D) at the early time of production since polymer itself has low injectivity as shown in Fig. 3 (red line).

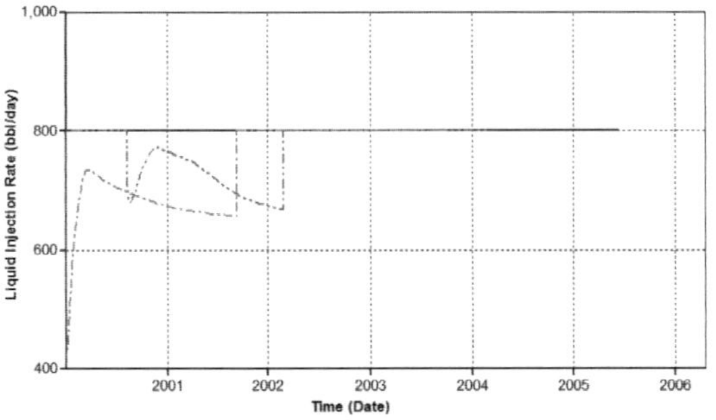

Fig. 3. Plot of liquid injection rate of polymer flooding (red), waterflooding (blue) and polymer flooding with pre-flushed water until water breakthrough (green) as a function of time

To increase injectivity of polymer, pre-flushed water is used to inject before a polymer slug. Several cases are simulated to determine which scenario will provide the best production scheme in terms of pre-flushed water and results are summarized in TABLE III.

TABLE III. RESULTS FROM DIFFERENT POLYMER INJECTION STARTING TIME

No.	*Scenario*	*RF (%)*	*Time (days)*
1	No Pre-flush	51.66	1,308
2	Pre-flush 25% before BT	52.89	1,492
3	Pre-flush 50% before BT	52.71	1,461
4	Pre-flush until BT	53.04	1,521
5	Pre-flush until 25% watercut	50.66	1,521
6	Pre-flush until 50% watercut	52.67	1,609

From the table, to inject pre-flushed water until water breakthrough (case 4) provides the best result since polymer injection rate can mostly attain desired rate at 800 STB/D after pre-flushed water is injected as illustrated by green line in Fig. 3. If less amount of pre-flushed water is used, injected water is not sufficient to assist increasing polymer injectivity. On the other hand, if pre-flushed water is kept injecting after water breakthrough, polymer injection time will be delayed and so does the ability to improve oil displacement from polymer. Hence, to inject pre-flushed water until water breakthrough is a trade-off between injectivity and effective oil displacement mechanism.

C. Alternating Water Slug Size

Another thing that can improve polymer injectivity is to split polymer slug into smaller ones and separate them with water slug in between. Polymer slug is cut in half (0.125 PV each) to perform double-slug polymer flooding while still remaining same polymer dosage. Alternating water slug size from 5%, 10%, 25%, 50%, 100% of polymer slug size is numerical simulated to determine the best scenario as illustrated in Fig. 4.

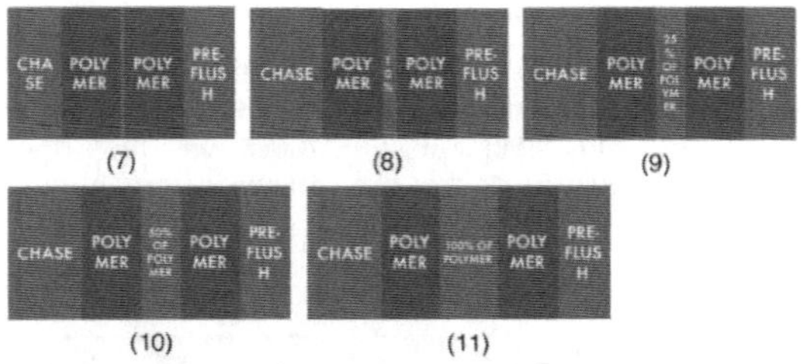

Fig. 4. Illustration of double-slug polymer flooding with different alternating water slug size

TABLE IV summarizes results in terms of oil recovery factor and total production time obtained from cases illustrated in Fig. 4. From the table, it shows that using the same alternating water slug size as polymer (case 11) provides the smallest oil recovery while using the smallest water slug at 5% of polymer (case 7) contributes to the highest oil recovery. This can be explained that even though it is true that if large water slug is selected, injection rate when injecting a following polymer slug will not drop much from the desired rate (800 STB/D) and it takes shorter time to inject last polymer slug but time spent during water slug injection is too much whereas the polymer injectivity from both cases is not significantly different as shown in Fig. 5. This setbacks time to inject another polymer slug. Another reason is that small amount of water is sufficient to desorb polymer on rock surface that has desorption level of 60%. In case that desorption level is more than 60%, more water may be required.

TABLE IV. RESULTS FROM DIFFERENT ALTERNATING WATER SLUG SIZE

No.	*Scenario*	*RF (%)*	*Time (days)*
7	Water : Polymer = 5%	53.69	1,613
8	Water : Polymer = 10%	53.11	1,521
9	Water : Polymer = 25%	53.08	1,552
10	Water : Polymer = 50%	53.17	1,576
11	Water : Polymer = 1:1 (100%)	51.31	1,674

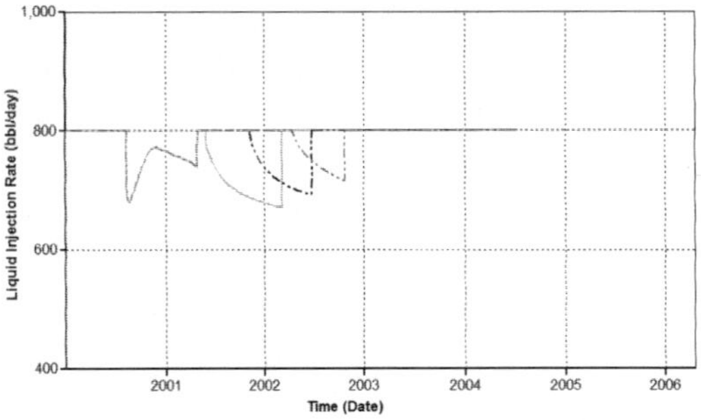

Fig. 5. Plot of liquid injection rate of double-slug polymer flooding using alternating water slug size 5% (green), 50% (black) and 100% (red) of polymer slug size as a function of time

D. Concentration Sorting

For double-slug polymer flooding in section C, concentration of each polymer slug is as same as that of bulk polymer slug. This section will focus on the effect of concentration order of double-slug polymer flooding both in descending (high to low) and ascending (low to high) order while keeping the same polymer mass as shown in Fig. 6.

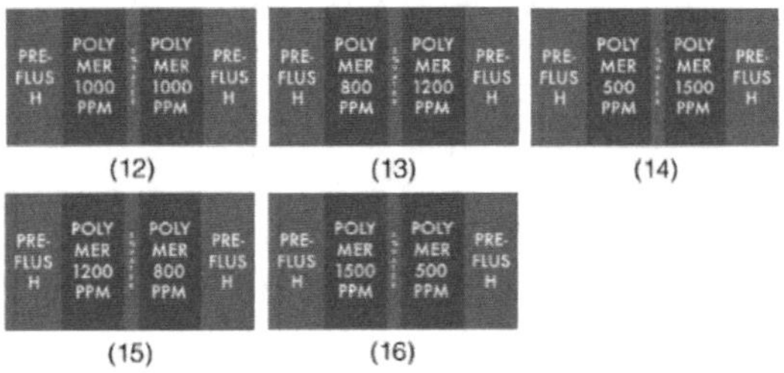

Fig. 6. Illustration of all scenarios of polymer concentration sorting

From TABLE V which summarizes results in this section, it is obvious that either descending or ascending order provide less oil recovery compared to case with the same concentration for all of the polymer slugs. This can be explained that for the first two descending order cases (case 13, 14), first polymer concentration is too high for pre-flushed water to recover the polymer injectivity while second polymer slug has less concentration that can be injected easily but with lower polymer effect so these greatly reduces the oil recovery afterwards. In ascending cases (case 15, 16), low concentration polymer is easily injected first so, the use of pre-flushed water to increase the polymer injectivity is unnecessary while following polymer slug has more concentration that can drastically reduce the injectivity of polymer since water slug in between is only 5% of polymer slug which is insufficient to increase the injectivity.

In conclusion, polymer concentration should be balanced for each polymer slug since it can obtain optimum effect from both polymer injectivity and effectiveness.

TABLE V. RESULTS FROM DIFFERENT CONCENTRATION SORTING

No.	*Scenario*	*RF (%)*	*Time (days)*
12	No Sorting	53.69	1,613
13	Descending (1,200/ 800 ppm)	51.44	1,582
14	Descending (1,500/ 500 ppm)	51.99	1,643
15	Ascending (800/ 1,200 ppm)	52.51	1,674
16	Ascending (500/ 1,500 ppm)	51.42	1,917

E. Number of Alternative Cycles

Since a bulk polymer slug can be segregated into several smaller slugs with water slug in between to increase polymer injectivity. The exact number of polymer slugs which provides the best result will be investigated in this section. A number of alternative cycles starting from 1 to 5 cycles as recommended by previous literature [5] are simulated with the same polymer mass and bulk polymer slug is segregated equally as illustrated in Fig. 7.

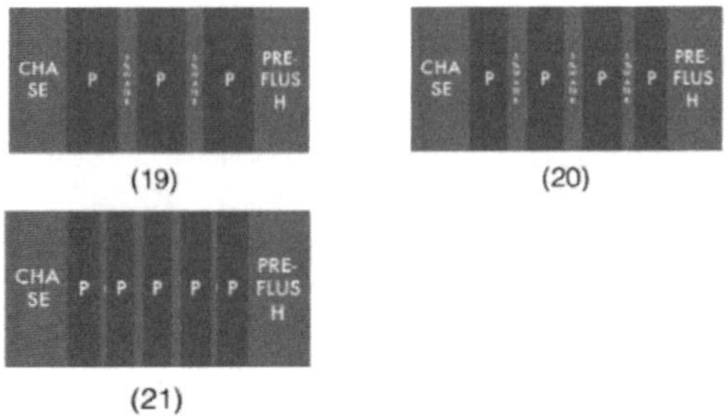

Fig. 7. Illustration of polymer alternating waterflooding with 3 cycles (19), 4 cycles (20), 5 cycles (21)

Results from TABLE VI shows that using 2 alternative cycles (case 18) contributes the highest oil recovery (slightly higher than using 3 cycles). This can be explained that the optimum alternative cycles for 0.25 PV, 1,000 ppm-polymer slug to be segregated is 2 cycles, if more than 2 cycles is used (case 19,20,21) , polymer dosage of each slug will be too small and not sufficient to maintain the effectiveness of polymer, resulting in relatively lower oil recovery.

TABLE VI. RESULTS FROM DIFFERENT NUMBER OF ALTERNATIVE CYCLES

No.	*Scenario*	*RF (%)*	*Time (days)*
17	1 cycle	53.04	1,521
18	2 cycles	53.69	1,613
19	3 cycles	53.66	1,613
20	4 cycles	53.56	1,582
21	5 cycles	53.61	1,582

Oil saturation profile for double-slug polymer flooding using pre-flushed water until breakthrough, water slug of 5% of polymer slug and same concentration for each polymer slug is illustrated in Fig. 8. The color scale shows different oil saturation starting from 0 (blue) to 1 (green). The injection sequence initiates from injecting pre-flushed water until water breakthroughs at producer (a) then, first 0.125 PV polymer slug is followed (b). After a short period of injecting water slug (c), second polymer slug is injected (d) and finally, waterflooding is performed until the end of production (e).

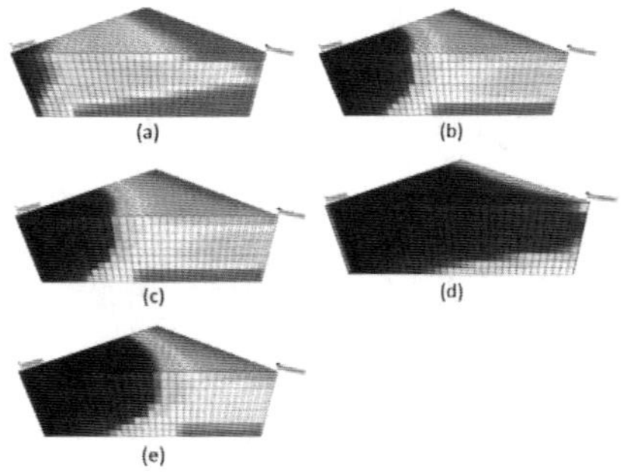

Fig. 8. Oil saturation profile of double-slug polymer flooding starting from pre-flushed water until breakthrough (a), first 0.125 PV polymer slug (b), water slug (5% of polymer slug) (c), second 0.125 PV polymer slug (d), chase water until the end of production (e)

V. CONCLUSION

1. Waterflooding causes early water breakthrough in heterogeneous reservoir so, polymer flooding is introduced to solve this problem since polymer can be served as a mobility controller for the reservoir. However, solely polymer injection can immensely reduce the injectivity. Consequently, *Polymer Alternating Waterflooding* is being studied to determine the best solution for this problem.

2. Pre-flushed water is not only used to check for the connectivity between wells but it is also injected to increase injectivity before injecting polymer slug. The study shows that to inject pre-flushed water until water breakthrough yields the best result.

3. Water slug between polymer slugs has ability to improve the ease of injecting another polymer slug and it is recommended to be relatively small in size compared to polymer slug. In this study, alternating water slug size of 5% of polymer slug contributes to the highest oil recovery.

4. Concentration of each polymer slug should be equal in order to balance the effect between polymer injectivity and oil displacement mechanism.

5. Two alternative cycles is the most appropriate cycles for 0.25 PV polymer with 1,000 ppm. Too many cycles means each slug will be too small and will not be effective to improve oil recovery.

6. In summary, injecting two polymer slugs should be performed with pre-flushed water until breakthrough, alternating water slug of 5% of polymer slug and same concentration for each polymer slug. This will provide oil recovery of 53.69% which is 0.65% (approximately 8,000 STB) higher compared to just injecting one single polymer slug with cumulative water production of 55,000 bbl over 1,613 days of production period.

ACKNOWLEDGMENT

The authors would like to thank to Chevron Thailand Exploration and Production, Ltd. for financial support for this study.

REFERENCES

[1] Al-Sharji, H.H., Grattoni, C.A., Dawe, R.A., Zimmerman, R.W., Disproportionate Permeability Reduction Due to Polymer Adsorption Entanglement. 2001, Society of Petroleum Engineers.Lake, L.W., Enhanced Oil Recovery. 1989: Prentice Hall.

[2] Lake, L.W., Enhanced oil recovery. 1989: Prentice Hall.

[3] Dominguez, J.G. and Willhite, G.P., Retention and Flow Characteristics of Polymer Solutions in Porous Media. 1977.

[4] Tiab, D. and Donaldson, E.C., Petrophysics: Theory and Practice of Measuring Reservoir Rock and Fluid Transport Properties. 1996: Gulf Publishing Company.

[5] Peihui, H., Haibo, L., Xu, H., Ruibo, C., Yuwei, Z., Xiaoyang, L., Alternative Injection and Its Seepage Mechanism of Polymer Flooding in Heterogeneous Reservoirs. 2015, Society of Petroleum Engineers.

 https://doi.org/10.1051/mateconf/201817104002

Effects of Formation Brine Compositions on Effectiveness of Low Salinity Brine Injection in Carbonate Reservoir

Suthida Meekangwal
Department of Mining and Petroleum Engineering
Faculty of Engineering, Chulalongkorn University
Bangkok, Thailand
Suthida.me@student.chula.ac.th

Falan Srisuriyachai
Department of Mining and Petroleum Engineering
Faculty of Engineering, Chulalongkorn University
Bangkok, Thailand
Falan.s@chula.ac.th

***Abstract*— Nowadays, Low Salinity Waterflooding (LSWF) is considered as one of the most effective techniques to enhance oil recovery. Essential mechanism is a Multi-Ion Exchange (MIE). In carbonate formation, collaboration of Potential Determining Ions (PDI) with rock surface and crude oil results in adjustment of surface equilibrium. These ions in injected brine are thoroughly investigated but effects from formation brine are not well documented. In this study, ratio of divalent ions including calcium and magnesium ions in formation brine is studied. Results from imbibition test shows that formation brine containing calcium ion to magnesium ion ratio of 1:1 yields response to LSWF. Oppositely, formation brines with high ratios of 10:1 and 15:1 do not respond to LSWF. As calcium ion in connate water diffuses to injected water, high calcium ion will exhibit dissolution mechanism of rock surface which is part of oil recovery mechanisms. Formation brine with high calcium ion is therefore not recommended for LSWF. High portion of magnesium ion in formation brine does not affect effectiveness of LSWF. Dissolution of magnesium from rock may be obstructed but oil can still be liberated from calcium carboxylate complex and dissolution of calcium. Coreflood test shows similar result which confirms previous explanation.**

Keywords—Enhanced oil recovery; low salinity water flooding; Carbonate reservoir; Multi-ion exchange mechanism

I. INTRODUCTION

Oil production can be generally performed by means of three recovery methods consisting of primary, secondary, and tertiary recoveries. During the phase of primary recovery, oil is produced by means of natural forces stored inside the reservoir, whereas in secondary recovery, conventional waterfooding and immiscible gas injection are performed to maintain reservoir pressure in order to prolong the production time. Nevertheless, most reservoirs have been foreseen to encounter difficulties of recovering oil by means of previously mentioned techniques. Thus, tertiary recovery is usually selected to perform in order to obtain extension of oil production. This recovery technique is implemented by injecting substances which are absent in the reservoir to generate oil recovery mechanisms beyond pressure maintenance or physical displacement process. In the past decades, waterflooding has been considered as one of the most effective methods for improving oil recovery due to availability of water as well as low investment cost and it is also environmental concerned process. Moreover, reservoirs containing moderate to light crude oil also exhibit positive results due to favorability of mobility ratio. However, complexity of water-oil-rock system sometimes causes waterflooding unable to exhibit the maximum capacity of oil production until residual oil saturation is achieved. This indicates that the extension beyond waterflooding is necessary. In recent years, one of the most attractive modified techniques of waterflooding so-called Low Salinity Waterflooding (LSWF) is investigated. This technique has been proven to recover more oil compared to conventional high salinity waterflooding. Furthermore, LSWF is a low cost technique since low salinity water could be prepared using seawater or fresh water mixed with formation brine without any additional chemical which can be harmful to subsurface environment.

Principally, LSWF is performed by injecting water containing low salinity or low ionic strength compared to that of formation water to create interaction onto rock surface. The cooperation between rock surface, crude oil, and injected low salinity water yields new surface equilibrium and hence; wetting condition is shifted toward a more favorable condition for oil production. LSWF is numerously studied in sandstone reservoirs and many studies have shown satisfactory. Several oil recovery mechanisms for sandstone reservoir were suggested such as fine migration, increase of pH and in-situ saponification, Multi-Ion Exchange (MIE), mineral dissolution, and Double Layer Expansion (DLE).

Nevertheless, carbonate reservoirs which contribute more than half of the world's oil reserves, have been rarely investigated for application of LSWF. From current studies, wettability alteration of carbonate rock surface is the most significant mechanism occurred during LSWF. Wettability can be changed due to a presence of potential wettability modifiers comprising sulfate ion (SO_4^{2-}), calcium ion (Ca^{2+})

and magnesium ion (Mg^{2+}) which are necessary factors for the occurrence of Multi-ion Exchange (MIE). This mechanism results in a more water-wet surface which is favorable condition for oil production. Nevertheless, most studies of LSWF in carbonate reservoir only focus on compositions of injected brine as a major cause of wettability alteration.

As part of surface system, formation water should be thoroughly investigated since it could control several mechanisms and it could be essential factor affecting the efficiency of LSWF as wettability alteration is a mechanism related to rock surface, injected brine, formation brine and crude oil. In this study, different ratios of divalent ions (calcium and magnesium ion) of synthetic formation brine are studied in carbonate rock by means of imbibition test and coreflood test to assess the impact of these Potential Determining Ions (PDI) for carbonate formations. Understanding effects of divalent ions in formation brine will be useful as it can screen candidate carbonate reservoirs that have high potentiality for LSWF technique.

II. THEORY

Low salinity waterflooding is a modified technique of waterflooding. The low ionic strength brine is injected into reservoir to create Multi-Ion Exchange (MIE) mechanism at rock surface and alteration of rock wettability to more favorable conditions for oil production is expected. A few studies have been performed on carbonate rocks and it is observed that oil recovery mechanism is occurred through collaboration of Potential Determining Ions (PDI) which consist of calcium ion (Ca^{2+}), magnesium ion (Mg^{2+}) and sulfate ion (SO_4^{2-}) [1]. Sulfate ion, possessing two negative charges, in injected water will compete with adsorbed organic acid in oil to adhere onto positively charged carbonate surface, causing less attraction force between oil and carbonate surface. At the same time, calcium and magnesium ions attract polar compounds in oil to form calcium or magnesium carboxylate complexes [5].

Nevertheless, appropriate number of calcium and magnesium ions should be presented in brine phase as reduction of these two ions also favors dissolution mechanism of calcium and magnesium (in case of dolomite formation) from rock surface to brine solution. It is also observed that dissolution of calcium ion results in increment of pH value due to accumulation of hydroxide ion (OH-) and as a consequence, carbonate surface may change its charge from positive to negative once the pH value rises beyond 9 which is point of zero charge of carbonate surface. When oil is liberated from rock surface, wettability is altered from oil-wet to a more water-wet direction and at the same time, oil recovery is increased. Fig. 1 illustrates oil recovery mechanisms by means of PDI in carbonate reservoirs.

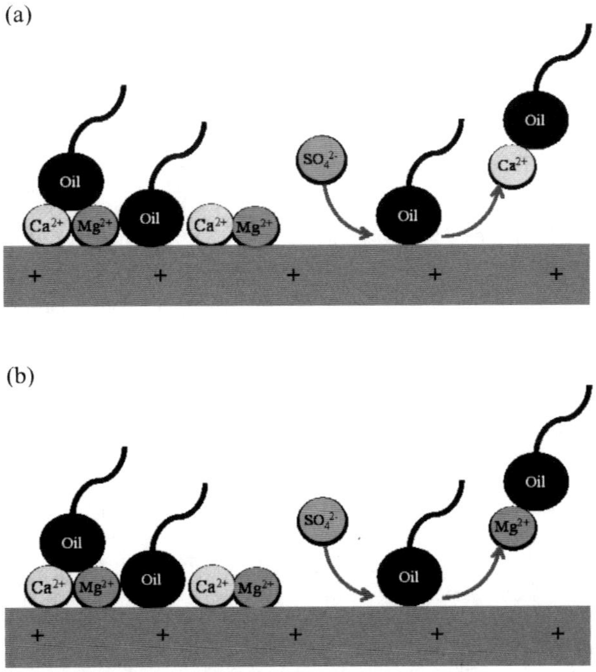

Fig. 1. Oil recovery mechanisms by means of collaboration between calcium ion, magnesium ion, and sulphate ion in carbonate reservoir, (a) Correlation between calcium ion and sulfate ion at normal temperature. (b) Correlation between magnesium ion and sulfate ion at high temperature.

Under reservoir conditions, formation brine, crude oil and rock surface are in equilibrium. Wettability is established from attachment of oil onto rock surface due to charge effect. Presences of different ions in formation brine therefore, can result in different strength of wettability and this can affect the effectiveness of several processes to enhance oil recovery including LSWF. In carbonate reservoir, total salinity as well as proportion of each ion can be varied. As divalent ions are found to determine effectiveness of LSWF, ranges of these two ions in formation brine are investigated in this study. From table I, calcium and magnesium ratio in carbonate formation is varied in the range of 3:1 to 15:1 and sulphate ion is mostly absent. Different ionic properties of calcium and magnesium ions such as ionic radius and ionic mobility may be differed from case to case as ratio of calcium to magnesium ion is changed and this could result in strength of wettability as well as amount of active ions during the LSWF.

III. METHODOLOGY

The experimental study is divided into two parts which are imbibition test and coreflood test. The formation brine compositions are varied in each test as an investigated parameter.

A. Determination of basic parameters and preparation of brines

Outcrop Silurian dolomite core samples with permeability between 100-400 mD and porosity between 0.16-0.18 are used in this study. Formation brine is prepared to have total salinity of 160,000 ppm which is obtained from the average total salinity of formation brine in carbonate formation around the world. Total concentration of divalent ion is fixed at 15,000 ppm. Seawater with total salinity of 35,000 ppm is used to represent low salinity water.

B. Performing of imbibition test to select appropriate ratios of divalent ions in formation brine

Core samples with diameter of 3.8 cm and length of about 6.5 cm are saturated with formation brine with various calcium ion to magnesium ion ratios in the range of 1:1, 3:1, 10:1 and 15:1 (Table I) by using coreflood apparatus followed by saturating core samples with dodecane which represents crude oil. Then, all core samples are soaked in formation brines for imbibition test and weighted every 24 hours to collect mass data for calculation of fluid saturations. After the weight of core samples is constant, all core samples are switched to soak in seawater to perform LSWF. Samples with the largest improvement in terms of oil recovery factor are chosen for displacement by coreflood test.

TABLE I. COMPOSITION OF FORMATION BRINE AND SEAWATER AROUND THE WORLD

Ionic species (ppm)	Reference [2]		Reference [3]		Reference [4]	
	FW	*SW*	*FW*	*SW*	*FW*	*SW*
Na^+	49933	13700	10350	23000	51820	10345
Ca^+	14501	521	1160	520	15992	521
Mg^{2+}	3248	1620	192	1080	1282	1094
K^+	0	0	195	390	0	392
Cl^-	111810	24468	18638	37985	111717	18719
SO_4^{2-}	234	3310	0	2304	0	2305
HCO_3^-	0	0	567	126	391	0
TDS	179730	43619	62800	33390	181202	33375

C. Performing of coreflood test to study effects of formation brine compositions to oil recovery performance

In this section, cases with improvement by means of LSWF from imbibition tests are selected to perform by using coreflood apparatus (fig. 2). Core samples used in step 2 are saturated with chosen formation brine followed by dodecane until residual oil saturation is attained. Saturated core samples are aged for 1 week and after that, core samples are flooded by formation brine to perform conventional waterflooding. Once amount of displaced oil is constant, seawater is injected to perform LSWF. Confining pressure of 2,000 psi and temperature of 30°C are applied to the system, whereas injection rate is fixed at constant value of 0.5 cm^3/min.

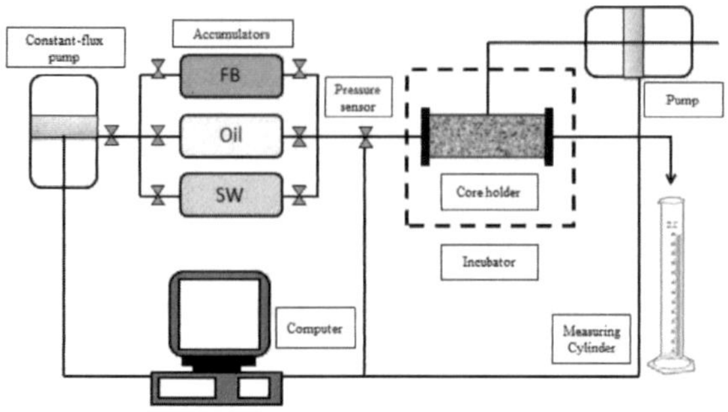

Fig. 2. Coreflooding system

IV. RESULTS AND DISCUSSION

Results obtained from imbibition test are illustrated in terms of relationship between oil recovery factor and time in fig. 3. As water and oil are different in density, amount of oil that is displaced can be calculated from the mass balance of total fluid weight in core sample. As water imbibes into core, expelling oil out, weight of sample is increased due to higher density of water compared to oil. From the figure, it can be noticed that oil recovery factor by means of formation water itself is quite small in most case except one with the calcium to magnesium ratio in brine of 1:1. When samples are switched to soak in seawater which is at the 8th day where oil recovery factors of all cases are constant, it is even more obvious that the case with brine containing calcium ion to magnesium ion ratio of 1:1 obtains the highest improvement in terms of oil recovery. At the end of the test, oil recovery factor of 26.02% is obtained. From the figure it can also be observed that the higher amount of calcium ion in formation brine, oil is difficultly displaced by both formation brine itself as well as seawater which is low salinity water. This can explain behaviors of calcium ion and magnesium ion during

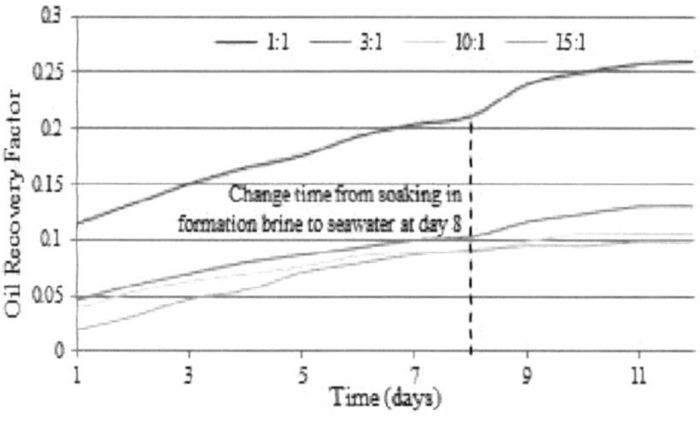

Fig. 3. Oil recovery factor from imbibition tests for formation brines with various ratios of calcium ion to magnesium ion as a function of time.

Presence of both calcium ion and magnesium ion is important for dolomite formation since dolomite composes of both ions inside its crystalline structure. Dissolution mechanism of both ions therefore depends on quantity of them inside enclosed water. Considering calcium ion first, calcium is one of the active ion in oil recovery mechanism by means of LSWF. Presence of calcium ion in injected water will form calcium carboxylate complex with adsorbed carboxylic acid. Attraction between oil and carbonate surface is then lessened and oil can be easily liberated. However, liberation of oil might require dissolution of calcium ion at the surface. This mechanism can occur rapidly when amount of calcium ion in injected brine is small. Therefore, dissolution can occur to adjust the equilibrium. When formation water contains high concentration of calcium ion, these ions from connate water will diffuse to injected water and increase total amount of calcium ion in the environment quickly. Even though calcium carboxylate complex can be formed, dissolution of calcium ion is obstructed and oil cannot be liberated.

Considering effects of magnesium ion, it is usually found that amount of magnesium ion in formation brine is less than calcium ion. Magnesium ion is also part of dolomite formation and similar to calcium ion, dissolution of magnesium ion may occur when low salinity water is injected. Nevertheless, magnesium ion cannot form magnesium carboxylate complex at low temperature and presence of magnesium ion in the environment does not cause much effect in oil recovery mechanism. Higher portion of magnesium ion can still prevent the dissolution mechanism of magnesium in formation. However, when portion of magnesium is high, portion of calcium ion is small and this favors dissolution of calcium ion in rock formation and amount of calcium ion in injected brine is adequate to form calcium carboxylate complex.

From imbibition tests, there are two cases where effects of LSWF is observed which are ratios of calcium ion to magnesium ion in formation brine of 1:1 and 3:1 and these two cases are chosen to perform coreflood test. Results are illustrated in fig. 4 and they still follow similar trend as obtained from imbibition test.

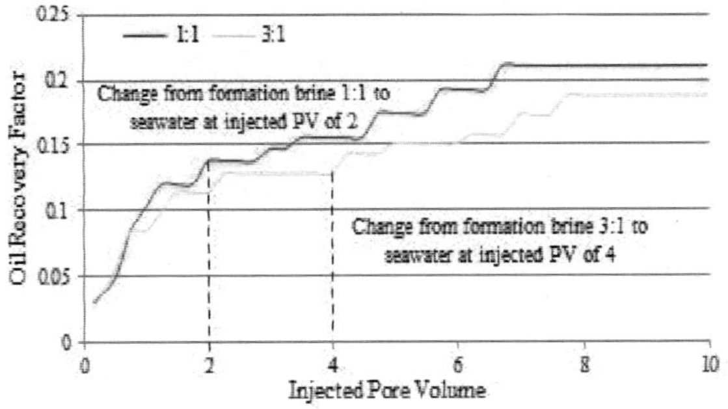

Fig. 4. Oil recovery factor from coreflood test with formation brine with various ratios of calcium ion to magnesium ion as a function of injected PV.

V. Conclusions

From this study, it can be concluded that high calcium ion concentration in formation brine is not favorable condition for LSWF in carbonate formation. As high calcium ion concentration in connate water can diffuse to injected water, environment with high amount of calcium ion starts to obstruct rock dissolution which is one of the favoring mechanisms for oil production. Amount of magnesium ion does affect the effectiveness of LSWF. Even though dissolution of magnesium ion may be obstructed from its high concentration, calcium carboxylate complex can be formed from adequate amount of calcium ion.

Acknowledgment

The authors would like to thank Chevron Thailand Exploration and Production, Ltd. for financial support of this study.

References

[1] T. Austad, S. Strand, M.V. Madland, T. Puntervold, and R.I. Korsnes, "Seawater in chalk: an EOR and compaction fluid," in press

[2] S. Chandrasekhar and K.K. Mohanty, "Wettability alterations with brine composition in high temperature carbonate reservoirs," in press.

[3] S.J. Fathi, T. Austad, and S. Strand, "Smart Water as a wettability modifier in chalk: the effect of salinity and ionic composition," Vol. 24, Energy Fuels, 2010, , pp. 2514-2519.

[4] R. Gupta, P.G.J. Smith, L. Hu, T.W. Willingham, M.L. Cascio, J.J. Shyeh, and C.R. Harris [2011] "Enhanced waterflood for middle east carbonate cores – Impact of injection water composition," In press.

[5] P. Zhang, M.T. Tweheyo, and T. Austad, "Wettability alteration and improved oil recovery by spontaneous imbibition of seawater into chalk: Impact of the potential determining ions Ca^{2+}, Mg^{2+}, and SO_4^{2-}," Vol. 301, Colloids and Surfaces A: Physicochemical and engineering aspects, 2007, pp. 199–208.

Carbon Dioxide Emissions Monitoring in Romania in the Context of Greenhouse Gases Reduction

Avram Lazăr, Cristescu Tudora, Stoica Monica Emanuela
Drilling, Extraction and Transport of Hydrocarbons Department
Petroleum - Gas University of Ploiesti
Ploiesti, Romania
avram_lazar55@yahoo.com, cristescuteo@gmail.com, monicastoica20022002@yahoo.com

Abstract— **The weather changes we are currently witnessing, characterised by dynamism and extreme phenomena, are the direct and indirect result of human activities which are determining the global atmosphere change in composition. The paperwork follows the evaluation and comparison of carbon dioxide emission coefficient in case of liquid oil fuels burning.**

The calculus model used for the carbon dioxide emission coefficient evaluation has been developed based on mathematical models from specific publications. It was applied in the case study based on data from specific literature. The results obtained following the evaluation allowed certain comparisons in the field of carbon dioxide emissions in case of the complete burning of certain fluid hydrocarbons use in industry.

Keywords— carbon dioxide emission coefficient, oil, monitoring

I. INTRODUCTION

In Romania, the negative effects of extreme meteorological phenomena generated significant economic losses in agriculture and rural development, transportation, energy supplying and water management.

The greenhouse gases are: carbon dioxide, methane, nitrogen protoxide, hydrofluorinecarbons, petrofluorinecarbons and sulfide hexafluorine.

The greenhouse effect is determined by these gasses' selective absorption of the Earth's thermal radiation and its isotopic reemission in the extra atmospheric space and towards the Earth. The infrared radiation reissued towards the Earth contributes to the low atmosphere warming, including the Earth itself [9].

Due to the increasing concentrations of these gases, the greenhouse effect intensifies and the energy and system moisture transportation is disturbed, leading to imbalances in the climate system.

A 4° C increase in atmospheric temperature by the end of the century would have a considerable impact on climate change, which would produce a considerable damage in Romania. Consequently, GHG emission reduction and adaptation to climate changes are priorities for Romania [9].

Romania was the first country to ratify the Kyoto Protocol (KP) and has reduced GHG emissions by 50% by the end of the period 2008 - 2012, although Romania's contribution to global emissions is insignificant (only 0.3% of emissions of greenhouse world and less than 3% of total emissions from EU countries) [9].

By sectors of activity, GHG emissions in 2011, are presented according to [8] in figure 1, the largest producer of the emissions being the energy sector.

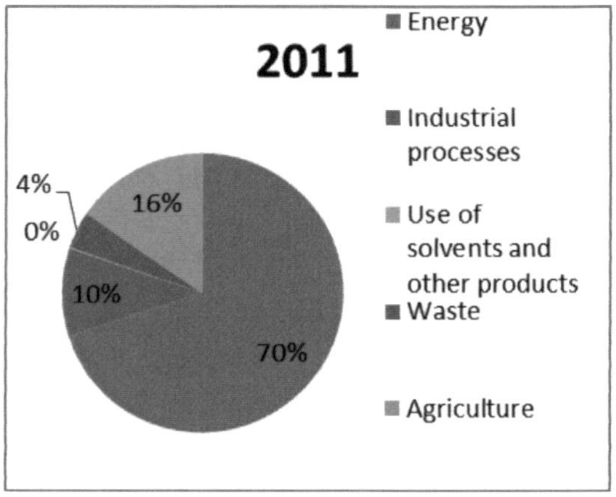

Fig. 1. Greenhouse emissions by field of activity

By 2020, Romania will have also to meet the targets of "20-20-20" within the "Energy & Climate Package" of the EU (a 20% reduction in emissions of GHG compared to 1990 levels; increasing the share of energy from renewable sources by 20% and 20% improvement in energy efficiency).

Monitoring of emissions [8] allowed establishing the intensity of GHG emissions in energy consumption for the period 2000 to 2013 regarding Romania (fig. 2), and also the CO_2 emissions' variations in intensity for the energy sector in 2000-2013 (fig. 3).

In conclusion, Romania proposed a strategy entitled "National Strategy on Climate Changes and Low Carbon Based Economic Growth".

This Strategy is developed in the context of a national vision, which must establish an economy resilient to climate change, with low carbon emissions, that will include policies and actions related to climate change, given the projected economic growth by 2050.

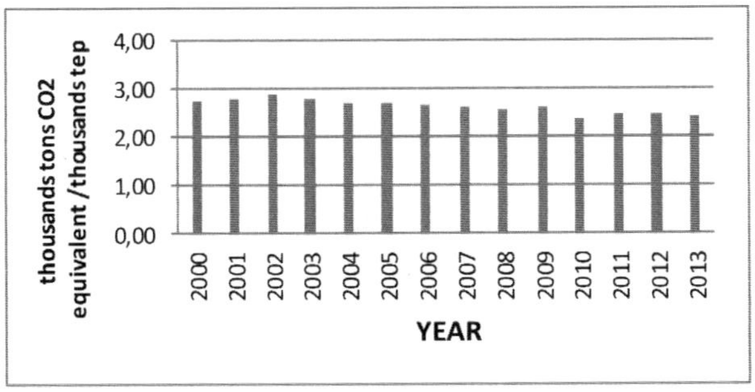

Fig. 2 Greenhouse gasses' variation in intensity for the energy consumption

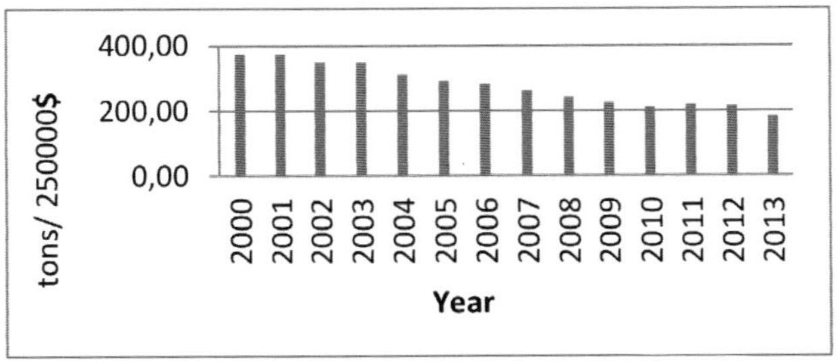

Fig. 3. CO_2 emissions intensity in the energetic field

Periodical statistical surveys on the production of electrical and thermal energy, resources and consumption of electrical energy, thermal energy, fuels, fuels and lubricants, monitoring of emissions of carbon dioxide lead to a classification using various criteria [2, 6, 7, 8].

According to the the label of electricity to end customers at regulated tariffs, (Table 1), electricity supplied by Enel Energie Muntenia SA in 2015 was produced from the following sources: hydroelectric, nuclear, natural gas, coal, fuel oil, wind, biomass, solar, unconventional sources and renewable [10, 11].

For 2015, total CO_2 emissions in Romania was 299.02 g / kWh of CO_2 emissions resulting from electricity generation were 32.53 g / kWh. It is noted that the environmental impact is below the national average.

TABLE 1. ELECTRICITY PRODUCTION SOURCES [11]

The primary source of electricity	Supplier ultimately, %	Romania electricity production in 2015, %
Coal	2.93	26.89
Nuclear	33.25	17.83
Natural gas	1.47	13.52
Fuel oil	0.01	0.06
Other conventional sources	0.02	0.16
Renewable sources	62.33	51.54
Hydroelectric	60.79	27.36
Eolian	1.20	11.03
Biomass	0.08	0.71
Solar	0.26	2.43

Carbon dioxide, CO_2, comes from flue gases resulting from combustion. A fuel must meet a number of conditions, namely, can be easily obtained from nature, is not toxic, is cheap and sufficient, so the cost of heat produced to be competitive at the time, on the energy market [3].

In Table 2 shows the development of the activities of the CO_2 emission energy, namely combustion plants and refineries [12].

TABEL 2. CO_2 EMISSIONS FROM THE ENERGY SECTOR ACTIVITIES, MILLION TONS [12]

CO_2 Emissions	2009	2010	2011	2012
Total CO_2 emissions for the energy sector, including refining	63.56	64.62	66.71	68.13
a. Installations for the production of electricity and heat	50.11	51.0	51.8	53.12
b. Installations for heat production for industry	5.5	5.63	5.85	6.0
c. Other plants for production of thermal energy in the residential, commercial, etc.	0.33	0.33	0.34	0.35
d. Refineries	7.62	7.66	8.72	8.66

II. EXPERIMENTAL PART

In the case of liquid petroleum fuels, the relative density of which is known, ρ_{15}^{15}, the fraction of carbon and hydrogen fraction can be calculated with the relationship [1]:

$$g_C = 0.15\rho_{15}^{15} + 0.74;\ g_H = 1 - g_C \qquad (1)$$

The heat issued following a complete combustion with the minimum necessary air, of a unit amount of fuel is called caloric power and is a fuels characteristic. Depending on the state of aggregation for the water resulting from the combustion, the following are known:

- high caloric power H_s - when water vapor contained in the flue gases are discharged as liquid;
- low caloric power H_i - where flue gases are discharged with water vapors in the gas.

To compute the caloric power of liquid petroleum fuels one can use the following empirical relationship [1]:

$$H_i = 46\,434 + 3169\rho_{15}^{15} - 8793\left(\rho_{15}^{15}\right)^2 \qquad (2)$$

Applying relation (2) leads to results expressed in kJ / kg.

The carbon dioxide emission coefficient, is used to compare the effects of combustion pollution. It is defined as the ratio between the mass of carbon dioxide resulting from combustion and the fuel's low or high caloric power. The relationship is:

$$\varepsilon_{CO_2} = \frac{m_{CO_2}}{H} \qquad (3)$$

The measuring unit for carbon dioxide emission ε_{CO_2} is expressed in g/kWh depending on the reporting, either using the high caloric power or the low one noted $\varepsilon_{CO_2,S}$ and $\varepsilon_{CO_2,I}$.

If the case of liquid petroleum fuels combustion, the carbon dioxide emission coefficient, considering that they have the composition only hydrogen and carbon can be calculated using the equations (1), (2) and (3) in the works [4] [5] computed as follows:

- CO_2 emissions, relative to the lower caloric power:

$$\varepsilon_{CO_2,I} = \frac{2714 + 550\rho_{15}^{15}}{12.898 + 0.88\rho_{15}^{15} - 2.442\left(\rho_{15}^{15}\right)^2} \quad (4)$$

- CO_2 emissions, relative to the higher caloric power: for liquid fuels not containing water:

$$\varepsilon_{CO_2,S} = \frac{2714 + 550\rho_{15}^{15}}{14.529 - 0.061\rho_{15}^{15} - 2.442\left(\rho_{15}^{15}\right)^2} \quad (5)$$

III. Results

Based on the mathematical model earlier depicted, the carbon dioxide emission coefficient was calculated for three types of oil, two domestic and one imported, for which densities are known (Table 3).

TABLE 3. OIL TYPE

Oil type	Relative density (the density of the liquid petroleum fuels relative to water, at 15 ^{0} C), ρ_{15}^{15}
Local light crude	0.8475
Local medium crude	0.8865
Impoted crude	0.8819

The obtained results are depicted in Table 4.

TABLE 4. CARBON DIOXIDE EMISSION BY COMBUSTION OF SOME LIQUEFIED PETROLEUM PRODUCTS [2]

Crude type	Carbon dioxide emission relative to the inferior caloric power [g CO_2/kWh]	Carbon dioxide emission relative to the superior caloric power [g CO_2/kWh]
Local light crude	267.46	250.00
Local medium crude	272.24	255.10
Impoted crude	271.80	254.40

Carbon dioxide emission analysis (Table 4) from burning hydrocarbon fluids leads to the conclusion:

$$\varepsilon_{CO_2,I,local\ light\ crude} < \varepsilon_{CO_2,I,local\ medium\ crude} < \varepsilon_{CO_2,I,imported\ crude}$$

The lowest values of the coefficient of carbon dioxide emission are related to the combustion of hydrocarbon fluids of low density.

IV. Conclusion

Research conducted in the paper used data on three types of crude oil.

From a qualitative perspective, the evaluation of the carbon dioxide emission coefficient provides a criterion for the fuel selection. Fuel with lower relative density is less polluting, whether domestic or imported.

From a quantitative perspective, knowing the amount of carbon dioxide emitted from burning of fuel allows to find the usage polluting effects.

A purchase selection criterion for a fuel, can be the carbon dioxide release.

The pollution effects quantification in case of liquid petroleum combustion can be achieved by evaluating the carbon dioxide emission coefficient.

REMARKS

g_i - Mass fraction of component i in the mixture and component kg / kg fuel

H - Caloric power, kJ /kg, kJ /m^3_N ;

m - Masa mass, kg;

ε_{CO_2} - carbon dioxide emission coefficient g/kWh;

ρ_{15}^{15} - Relative density (density of an oil product at 15 ^{0} C, relative to the density of water at 15 ^{0}C)

SYMBOLS

C- Carbon
CO_2- Carbon dioxide
H- Hydrogen
I- Low
S- High
GHG - greenhouse gases

References

[1] A.R. Cernea, D. Dobrinescu, I. Fagarasanu, A. Covaci, Thermotechnics, Ed. Diactica and Pedagogica, Bucuresti, p. 273, 1969.

[2] T. Cristescu, M. E. Stoica, Evaluation of the Emission of Carbon Dioxide in the Combustion of Biofuels and Comparison with Liquid Petroleum Products, Indian Journal of Applied Researche, vol 3, p. 28-30, October 2013.

[3] T. Cristescu, Termotechnics, Ed. University of Ploiesti, Ploieşti, p. 145, 2009.

[4] T. Cristescu, Liquefied Natural Gases and Liquefied Petroleum Gases as Complementary Sources of Energy, 5th International Conference on Energy and Environment, Bucureşti, 3-4 November, 2011.

[5] T. Cristescu, The transport and storage of liquefied gas, Ed. University of Ploiesti, , p. 258 – 259, Ploieşti 2012.

[6] T. Cristescu, M.E. Stoica, GH. Brănoiu, T. Negreanu-Pirjol, Evaluation and Comparing of the Carbon Dioxide Emission Coefficients for the Combustion of Gaseous and Liquid Hydrocarbons, Revista de chimie (Bucharest), vol 65, p.856-860, July 2014.

[7] K. Raznjevic, Thermodynamic tables and diagrams, Ed. Tehnică, Bucureşti, 1978, p. 327.

[8] www.insse.ro Institutul Naţional de Statistică, România.

[9] www.mmediu.ro/app/webroot/.../2015-08 2020_CRESC_Strategia_climatica_V3.pdf.

[10] Ordin ANRE nr. 69/2009 – Regulament de etichetare a energiei electrice

[11] www.enel. ro/ro/ clienti-em.html.

[12] http://www-old.anpm.ro/files2/TEXT%20Anexe%20HG_NAP_ro-%20FINAL_20098183817246.pdf

AUTHOR INDEX